THERE'S NO SUCH THING
AS AN IT PROJECT

EX LIBRIS

HUDSON LIBRARY
& HISTORICAL SOCIETY

There's No Such Thing as an IT Project

A HANDBOOK FOR INTENTIONAL BUSINESS CHANGE

BOB LEWIS & DAVE KAISER

Berrett–Koehler Publishers, Inc.

Berrett-Koehler Publishers, Inc.
1333 Broadway, Suite 1000
Oakland, CA 94612-1921
Tel: (510) 817-2277
Fax: (510) 817-2278
www.bkconnection.com

ORDERING INFORMATION

Quantity sales. Special discounts are available on quantity purchases by corporations, associations, and others. For details, contact the "Special Sales Department" at the Berrett-Koehler address above.

Individual sales. Berrett-Koehler publications are available through most bookstores. They can also be ordered directly from Berrett-Koehler: Tel: (800) 929-2929; Fax: (802) 864-7626; www.bkconnection.com.

Orders for college textbook / course adoption use. Please contact Berrett-Koehler: Tel: (800) 929-2929; Fax: (802) 864-7626.

Distributed to the U.S. trade and internationally by Penguin Random House Publisher Services.

Berrett-Koehler and the BK logo are registered trademarks of Berrett-Koehler Publishers, Inc.

Printed in the United States of America

Berrett-Koehler books are printed on long-lasting acid-free paper. When it is available, we choose paper that has been manufactured by environmentally responsible processes. These may include using trees grown in sustainable forests, incorporating recycled paper, minimizing chlorine in bleaching, or recycling the energy produced at the paper mill.

Library of Congress Cataloging-in-Publication Data

Names: Lewis, Robert L., author. | Kaiser, Dave, author.
Title: There's no such thing as an IT project : a handbook for intentional business change / Bob Lewis, Dave Kaiser.
Other titles: There is no such thing as an IT project
Description: 1st Edition. | Oakland, CA : Berrett-Koehler Publishers, 2019.
Identifiers: LCCN 2019011716 | ISBN 9781523098835 (paperback)
Subjects: LCSH: Project management. | Information technology—Management. | Leadership. | Organizational change. | BISAC: TECHNOLOGY & ENGINEERING / Project Management. | BUSINESS & ECONOMICS / Industries / Computer Industry. | BUSINESS & ECONOMICS / Leadership.
Classification: LCC HD69.P75 L496 2019 | DDC 004.068/4—dc23
LC record available at https://lccn.loc.gov/2019011716

First Edition
25 24 23 22 21 20 19   10 9 8 7 6 5 4 3 2 1

Set in Palatino by Westchester Publishing Services.
Cover designer: Adam Johnson

To the many longtime readers of InfoWorld's
"IS Survival Guide" and its successor, my weekly
Keep the Joint Running *e-letter. They have, over the*
years, kept me motivated, honest, and most
important, better informed.

—Bob

To the second boss in my career, Dave Crowley,
who once told me, "Kaiser, if you ever stop complaining
you will have a great career." Dave was a great mentor
and taught me the importance of being open and
honest with those you care about.

—Dave

Contents

Foreword

Anita Cassidy

Partner at ITDirections (www.itdirections.com)
and author of five books, including *A Practical Guide
to Information Systems Strategic Planning.*

Why is this book important? It's because today, technology is at the core of all businesses. It is difficult to find a business strategy or business process that does not depend on technology for its success. In fact, for differentiating, innovating, or disrupting organizations, the business and the technology are typically inseparable. Just look at companies such as Uber, Airbnb, Amazon, Houzz, Netflix, Zipcar, Tesla, Instacart, Salesforce, Facebook, Google, or Apple; it is difficult to see the business and the technology as separate entities. The business is the technology, and the technology is the business.

For decades, IT practitioners have hammered on the idea that the business and IT need to be aligned. In today's world, with technology and the business essentially being one, alignment isn't enough anymore. As Bob and Dave point out, IT can't just be aligned with the business—it has to be integrated into it. Which is why many are coming to see that there is no such thing as an IT project. They're business-change projects that are using technology as an enabler.

Digital transformation is the current discussion and hype in the industry. What is Digital transformation? It is nothing more than business change, effectively using technology through-out all aspects and touchpoints of the business.

This book is important because it's about change. Compa-nies in all industries and of all sizes must continually change to be relevant and successful. Change must happen quickly, intentionally, and flawlessly. Yet, change of any kind is diffi-cult. Whether you are talking about changing a culture, a business, a business process, a toolset, a skill, a personal habit, or a mind-set, change is tough. There is no surefire solution or recipe for change.

This book provides excellent thoughts and advice for navi-gating a course of change in your organization. Change in every organization and every project is different. What may be successful in one organization or project could be disas-trous in another. This book provides many creative ideas and suggestions for business change.

Bob and Dave have a unique and entertaining writing style. Not only is their writing enjoyable, but they make you think. They make you question the obvious and the not so obvious. They provide original, concrete, and pragmatic suggestions and advice that you can put to immediate use. At the begin-ning of each chapter, they provide a real-life story, or interlude, to help make the material relevant. At the end of each chapter, they provide a succinct summary of the key points, titled "If You Remember Nothing Else," and make it actionable with a section titled "What You Can Do Right Now." Whether your job function is in the business or IT, I think you will find thought-provoking value and useful advice in this book that will help you navigate your unique path of, as Bob and Dave call it, *intentional* business change.

Prologue

What the Mess Is and How We Got into It

An analogy isn't the same thing as being the same thing.
—The Economist

There's a famous cartoon in IT circles[1] (figure 1) that shows a succession of swings, as specified in the project request, as explained by the systems analyst, as built by the programmers,

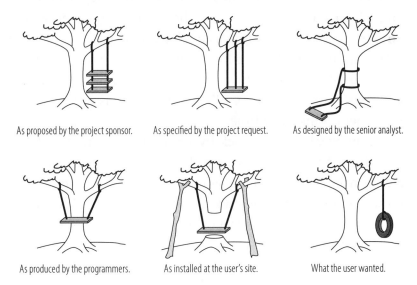

As proposed by the project sponsor. As specified by the project request. As designed by the senior analyst.

As produced by the programmers. As installed at the user's site. What the user wanted.

FIGURE 1 An application development metaphor.

and so on. None of them could actually work, and all of them look entirely ridiculous.

Except, that is, for the last panel of the cartoon, which shows what the users really wanted.

It shows a tire, hanging by a rope from a tree limb.

Only that doesn't tell the story as this book would tell it. Our view: what the users wanted wasn't the tire swing itself. The final panel of our cartoon would show something quite different.

It would show children playing on the swing, having fun.

Two Anecdotes

A rather excitable former client once became quite agitated when I (Bob Lewis) asked which of the six dimensions of optimization (explained in chapter 2 if you want to read ahead) were most important for a critical business process we were discussing.

"I don't have to choose!" he shouted (not an exaggeration). "I do Six Sigma! I can improve all of them at the same time, and I don't even need information technology to do it!"

He very well might have been right at that. If your business processes are bad enough you probably can improve them on all fronts at the same time. What you can't do is optimize for all of them at the same time.

In any event, I was struck by his proud assertion that he wouldn't need any information technology, because this has become an important talking point among business improvement methodologists. It's an attractive selling proposition, built around expectations that if you need information technology you're in for IT projects, which are notorious for being expensive at the start, with inevitable cost overruns later on if outright failure doesn't happen first.

Whatever the reason, management consultants have, for the most part, done everything they can to decouple business improvement from information technology delivery.

Meanwhile . . .

Once upon a time a decade or so ago there lived an online retailer. The company was quite successful—successful enough that its leaders were on a constant lookout for what the next level might look like.

Then they heard a presentation from a purveyor of eMerchandising software (call it EMS). The online retailer's decision-makers liked what they saw, signed a contract, and launched the EMS Project.

The company's resident strategic consultant* urged the project sponsor not to do this. It wasn't that EMS was inferior to its competitors or that installing it would be a bad idea.

It was the project's name. The EMS project was all about installing, configuring, and integrating software into the company's existing portfolio.

What the consultant unsuccessfully recommended was to call the effort the "Online Merchandising, Oh, by the Way, Using EMS" project instead.

The project was successful according to the Project Management Institute's definition of success: it was completed on time, with unchanged scope, and within its original budget.

And yet, a month after the project had finished, the consultant heard two of the company's senior web designers arguing about which of their new home page layouts was better.

"Why," asked the consultant, "don't you use EMS to A/B test them, to see which one drives more sales?"

"It does that?" they asked.

---

* Your loyal lead author again.

Why IT Projects Are a Bad Idea

You could stop reading right here and get value from this book, because this is the point of it: there are lots and lots of IT projects going on right now, all over the world.

That's the problem, because with few exceptions they're mistakes. It isn't that there is no such thing as an IT project. It's that there shouldn't be. Because if you're undertaking an IT project—if all you're going to do is install software—what you're going to deliver is shelfware.

Or, almost as bad, you'll deliver something reminiscent of an old Steven Wright joke: "I dreamed someone stole everything I owned and replaced it with an exact duplicate."

Only here what's going on isn't theft. It's far stranger than theft. What happens a lot is that companies buy new and expensive software to replace the old legacy systems they can't do without but that cost a lot and require the talents of programmers who are nearing or beyond retirement age to maintain.

And . . . this is the punchline . . . they do everything they can to make the shiny new software act just like an exact duplicate of the software they're retiring. Why? It's easier, less risky, and nowhere near as disruptive. Except that just about all of the business benefit comes from the disruption.

And so, after decades of this praxis, skeptical business-opinion influencers make a big fuss about how little return businesses get from their investments in information technology—opinions that demonstrate it's possible to be simultaneously accurate and completely wrong. Because often, businesses get too little return from their investments in information technology because they choose to get too little return.

See, the purveyors of these expensive software suites don't charge their sky-high license fees because their sales reps are extraordinarily persuasive, conning IT decision-makers who are too naive to figure out these are overpriced behemoths.

They cost as much as they do because they provide an enormous set of potential capabilities many businesses consciously decide to ignore.

They ignore them because, as you'll read in the pages that follow, the methodologies available to them . . . convincingly presented, so long as you consider the phrase "best practice" to be convincing . . . provide little or no guidance for how to use new information technology capabilities to improve how their employees get things done.

How did we get into this mess? The first two steps on the path to this particular circle of perdition were (1) drawing an analogy and (2) taking it seriously.

The analogy was that because IT organizations deliver technology to the rest of the business, they're "like" any other organization that delivers technology to someone else, which is to say, they're like software businesses.

And so they are, in that both they and software businesses create applications for someone else to use.

Interestingly enough, that's where the analogy ends, or should, not that you'd know this from reading the IT industry press.

It should end right there because software companies—think SAP, Oracle, Microsoft, Salesforce, and their brethren—create software products they sell to thousands and sometimes millions of business customers. When an IT organization writes software, in contrast, that software is designed for one business to use.

There are more than a few differences between having one metaphorical "customer" and having ten thousand real, paying ones. For example, with one customer you can tailor the software for your customer's exact situation. With ten thousand you can't.

On the other hand, with ten thousand customers you can afford to spend a lot more on your product. Beyond that, you're competing for business with other companies that sell similar products. Which means you can and have to spend a lot more designing and building your software package than any internal IT organization can or should afford to.

One more difference: Internal IT ought to, and usually does, have the best interests of the organization it serves at heart. Software vendors, in contrast, while they generally do want their customers to succeed, care more about making a sale than about what's truly in their customers' best interests.

Nonetheless, the analogizers were (and still are) persuasive. And so we're suffering from what so often happens with analogies: rather than recognizing where the analogy holds and where it breaks down, we've taken every characteristic of software businesses and insisted IT adopt identical practices.

In particular, IT's project methodologies almost universally start with the notion that IT's job is finished when the software "satisfies requirements" and "meets specifications." They finish with the proposition that business cost centers should pay IT for its products and services. Why? Because that's what businesses do: charge and pay for the products and services they provide and consume.

Except that when you run a business the point of the exercise is to sell products and services at a profit. When you run internal IT, the point of the exercise is, or rather should be, to collaborate in and provide support for intentional, beneficial business change.

How We Got into This Mess

Once upon a time, deep in the dim past when computers were first starting to expand beyond their initial accounting toehold in the world of business, programmers didn't understand their job was to deliver a product to their internal customers. For that matter, they didn't understand they weren't supposed to be capable of talking to their nontechnical counterparts elsewhere in the organization.

Quite the opposite. In the early days of IT, business managers were quite comfortable asking programmers if they could get the computer to do something or other to help them run their part of the business better, and programmers were equally comfortable either making it do thus-and-such or explaining that no, computers don't do that, but they could do some other thing that might improve the situation.

After a bunch of conversations like this, the programmer had the computer doing so much that it would have been hard for the business area that used these programs to go back to pencil-and-paper techniques.

Without even necessarily meaning to, the programmer and the business manager had built a big honkin' system that ran a significant chunk of the enterprise. These big honkin' systems vastly improved the business processes they touched, long before Lean, Six Sigma, Theory of Constraints, and Business Process Reengineering became part of our lexicon.

If they (and "they" includes "we") had only known, we and they would have written up "talking with each other a lot" as a methodology, called it "Agile," and made a fortune.

The big honkin' systems (we now call them "legacy systems") built through this series of can-you-get-the-computer-to-do-this conversations did sometimes get a bit messy. That's because neither the business manager who asked nor the programmer who delivered knew where things were going to end up. So they built a puzzle piece, then a second puzzle piece that fit into the first. The third piece had to fit together with the first two, and every additional piece increased the danger they'd paint themselves into a metaphorical corner.*

And so, a bunch of methodological theorists came along to make sure this never happened again. They managed this by tossing around another analogy that doesn't work very well: developing software, they explained, is exactly like building a skyscraper.

Or it should be. Because you never see skyscrapers with internal design inconsistencies.

This is how the so-called Waterfall methodologies came to be, with their legendarily high rates of failure and their usual mismatch between what the business needs and what projects built around Waterfall methodologies deliver.

Not that it matters from the perspective of this book, but among the many analogy breakdowns these fine theorists might have noticed but didn't is an important difference between computer software and skyscrapers: users of software but not of skyscrapers never run out of new requests for "Can you get the computer to do this too?"

What matters a lot from the perspective of this book is their definition of success, taken whole cloth and without much scrutiny from the Project Management Institute's skyscraper guidelines.

Success, according to Waterfall methodologies, means the project came in under budget, on schedule, and with all the originally planned features intact.

---

* Yes, the metaphor police are going to hunt me down like a dog.

Please understand—your loyal authors have nothing against a project coming in on time, staying within budget, and delivering what it's supposed to deliver.

We just object to calling this success.

Come to think of it, this isn't much good as a definition of skyscraper success either: when someone builds a skyscraper, it's a failure unless lots of people and businesses want to live or work in it.

Introduction

What You're in for When You Read This Book

*If you board the wrong train it is no use running along
the corridor in the other direction.*
—Dietrich Bonhoeffer

This is a book about achieving intentional business change.

It's also a book about why your company should never again charter an IT project. The two statements are opposite sides of the same coin.

In planning this book we had two alternatives. We could have taken one or more of the various business process improvement methodologies as our starting point, incorporating information technology change into them.

Or, we could have started with the current crop of IT methodologies, enhancing them so their purpose isn't to deliver software products anymore but to make business change happen, and happen as intended.

It was no contest. Here's why.

First and foremost, in most businesses IT takes project management (the discipline of making tomorrow different from yesterday) far more seriously than most of the rest of the enterprise. As a result, its methodologies are more detailed and mature than those devoted to business change, as are its practitioners.

Second and almost as foremost, the largest slice of most companies' business change investment pie chart is generally the spending needed to make the necessary changes and enhancements to the applications' portfolio.

And third, modern IT project methodologies are all about making IT projects better at adapting to changing conditions. The current crop of business process improvement methodologies are, in contrast, mostly about making businesses more efficient at what they did yesterday and will continue to do tomorrow.

So while this book is about achieving intentional business change, it starts with information technology and works its way out from there. It's a better place to start.

But just to start.

A Parable

IT implements and runs software. Business managers use the software to help them get their jobs done. IT supplies, the rest of the business consumes. What's the problem?

Here's one: by turning IT into a supplier of products and services used by the rest of the business, we create a covert conspiracy of failure. It works like this:

There's little that enhances a business executive's career more than envisioning and advocating bold, transformational change. And whether or not anyone wanted to admit it, most bold, transformational business change has, for the past several decades, required new or transformed information technology.*

And so the business executive becomes the project's business sponsor for either developing or buying, integrating, and configuring the new information technology.

Or, if it's an Agile project, the project's product owner.

Meanwhile, on the supply side of things, there's little that enhances IT's reputation better than developing slick, sophisticated software, with the likely exception of IT installing, integrating, and configuring a slick, sophisticated software package.

So far, so good. IT and the rest of the business are, as the pundits like to say, in alignment.

As it turns out, they're often in better alignment than anyone suspects, because while advocating bold transformational change is a career-enhancing move among business executives, actually pulling the trigger and making the change happen is a highly risky proposition.

Risky? Yes, risky, because every bold, transformational change constitutes a bet on what the future will look like, coupled with the hope that the analysis and design process has identified and planned for most ripple effects, and that any that were missed . . . the unintended consequences . . . won't be overly pernicious.

Guess right and the change makes the company more profitable. Guess wrong and every penny and erg spent

---

* Among those loathe to admit it are the practitioners of the various process optimization methodologies. Lean, Six Sigma, and Lean/Six Sigma in particular are replete with case studies (ManagementSpeak for "fairy tales") that purport to report transformations without ever involving IT. Usually, the stories turn out to be minor tweaks, not major transformations.

(depending on whether you're tabulating time or effort) have been wasted.

The executive who's advocated the bold, transformational change will, in many cases, turn out to be much happier if the actual change never takes place, even when the executive is the business sponsor and is a theoretical coconspirator.

So long, that is, as the executive has an ironclad not-my-fault excuse to fall back on.

How about IT? It's like this: programmers generally like to program. Programmers writing code are like painters applying oil to canvas: it's what they like to do, and getting paid for it is even better.

Until, that is, it's time to put their software into production. That's when programmer risk happens, because the software the programmers so lovingly build might not actually, for example, do anything useful.

And so it comes to pass that when programmers finish their software, the business executive explains there's no possible way employees can get their jobs done using what the programmers have wrought.

The programmers, politely but through clenched teeth, point out that the software satisfies all documented requirements and meets all specifications. They did their job and they did it right.

"I don't care about the requirements and specifications," the business executive replies. "I didn't understand them in the first place. I only signed off because I had to or IT would never have started programming."

Both IT and its so-called customer got what they wanted: they did what they were supposed to do, and it's the other one's fault nothing useful came of it.

A Less Cynical View

Okay, fair is fair, and our descriptions of how business executives and programmers think aren't. Fair, that is. Both parties deserve a bit more credit. While the conversations are often as we describe them, it's more likely the executives in question are experiencing cold feet than expressing their inner Machiavelli.

IT's developers, in their turn, are more often exhibiting the natural tendency all parents have when someone calls their baby ugly than coldly and analytically comparing executive complaints to the requirements and specifications documents.

And yet, the arguments we describe are more the rule than the exception. Business executives complaining that software is unusable as implemented are commonplace.

Programmers blaming the executives for not understanding their requirements, for signing off on specifications they didn't take the time to read and comprehend before doing so are just as commonplace.

IT's chronic failure to deliver solutions business stakeholders find delightful is very real, whether the mismatch between what they actually wanted and the requirements and specs as interpreted by the programmers was the result of intentional cynicism or honest error.

The real problem is what it is whenever the question is whose fault something is: blamestorming has taken precedence over root cause analysis.

And the root cause is right in front of everyone involved: even imputing the noblest of motives to all concerned, when it's an IT project the goal is to get the software right.

Not that the goal should be changed to getting the software wrong, although there are cynics in many businesses who might claim that's what's happened.

It's that when projects are about the information technology, what gets lost in the shuffle is the point of it all: the intended business change that will make one or more parts of the business run differently and better.

Which leads to unfortunate consequences, such as business stakeholders figuring they're doing IT a favor by sparing some of their staff's time and effort to serve as subject matter experts (SMEs) for a project. Which staff members' time and effort do they spare? The ones who have the most time available, not the ones best suited to making the project a success.

Sometimes it's even worse: because business logic is embedded in code and has been for a decade or more, it can be hard to find anyone who knows, with any level of depth, just exactly what that business logic is, and why. And yet, when deploying a replacement system, the affected business areas still need to take responsibility for the business logic to be embedded in the new system.

When there's no such thing as an IT project—when projects are defined in terms of the business outcomes they're supposed to bring about—here's what's different. It's all good, and it's the ground this book will cover:

- **It's Always the Culture:** Culture consists of shared assumptions, attitudes, and knowledge (we'll provide a more precise definition when we arrive at this section). For IT projects to give way to projects that deliver intentional business change, both business management and IT culture are in for some adjustments:
  - **Business management culture:** With IT projects, business management justifiably figures the project is Someone Else's Problem, which is an extension of most large enter-

prises' overall silo orientation. It's a culture predicated on the proposition that the world is divided into Things That Are My Problem and Someone Else's Problems.

To eliminate IT projects, the company's top executives have to replace, or at least discourage, silo thinking, strongly encouraging ubiquitous collaboration in its place.

- **IT culture:** Talk with your average IT manager or professional about their relationship with the rest of the business. What you'll hear . . . and even more significantly what you'll see in the way of expression and body language . . . is defensiveness. Because so much depends on information technology, when just about anything goes wrong it's IT that's left holding the bag.

    When IT is a supplier to internal customers, and projects are finished when it delivers the software, defensiveness reigns supreme in IT, followed closely by self-righteous aggression: "How are we supposed to deliver software they like when they don't know what they want?"

    We'll talk about changing the business management and IT cultures in more depth in chapter 1.

- **The New Business/IT Conversation:** The no-IT-projects transformation starts with a simple-sounding change: instead of asking a so-called internal customer, "What do you want the system to do?" IT's representatives will get in the habit of asking, "How do you want your part of the business to run differently and better?"

    Chapter 2 investigates all the different ways this changes everything for just about everyone else in the enterprise.

- **Fixing Agile:** When everyone stops defining project objectives in terms of software delivery, IT still has to either design and build or select, install, integrate, and configure business applications.

When that time comes, when it comes down to it there are two types of IT organizations: those that have embraced Agile and those with a high project failure rate.

Agile is a big honkin' deal.

But Agile as usually practiced is still about IT product delivery, not about intentional business change. As usually practiced it has other limitations as well.

If you aren't clear what the difference is between IT's Agile methodologies and the business being agile, no worries. We'll explain it all, and what to do about it, in chapter 3.

- **IT Operations:** So far we've talked only about the application development side of the IT house. It's time to let IT Operations in on the fun.

  IT Operations projects are a lot closer to being IT projects in that their purpose is to deliver working technology. In most cases they're more along the lines of upgrading platforms to current versions than on directly changing how parts of the business can run differently and better.

  But IT Operations doesn't generally deliver the results to business stakeholders so as to help the recipients run their parts of the business differently and better.

  From a business perspective these projects are more about risk avoidance and mitigation, which means that for the most part IT can keep them out of view, except for minor matters like budgeting the time and money needed to make them happen.

  But that doesn't let IT Operations entirely off the hook. On a day-to-day basis, IT Operations and the rest of the business are tied together even more closely than the applications side of IT. Chapter 4 covers this ground. Spoiler alert: the keyword is "BusOps."

- **Business-Change Governance:** When businesses have IT projects they have IT governance or IT project governance, and it's all about making sure IT doesn't just spend money implementing "technology for technology's sake," whatever that means.

    With business-change projects to govern instead of IT projects, there's clearly no longer a need for IT project governance. Instead, business-change governance will replace it. But if all that happens is assigning a new name to the same old practices . . . well, when has changing anything's name without changing its substance ever achieved anything important?

    Chapter 5 explains what has to change for business-change governance to be effective.

- **IT's Role in Business Planning:** When using computers in business was new and exciting, whoever headed IT spent a lot of time working with business executives and managers explaining what automation could do for them. IT played an active role in business planning, from strategy on down to specific automation opportunities.

    Then the business became IT's customer and all of that stopped. IT became an order taker. Which was much safer but far less satisfying.

    A key takeaway from the now-prevalent and, among skeptics, faddish conversations about Digital strategy and transformation is that IT needs to be more than a responder to business needs. Most new business opportunities and threats come in the form of newly available

technology, much of which is information technology.*
That being the case, IT should play an essential role in
formulating business strategies.

   Again.

   That's what chapter 6 is about.

- **The Seven Change Disciplines:** Achieving intentional
  business change calls for seven critical disciplines: leadership,
  business design, technical architecture management, application development, organizational change management,
  implementation logistics, and project management. A business that masters these will turn its strategies into business
  reality.

   The final chapter (7) provides a quick sketch of each of
   them. Business leaders who want their organizations to
   achieve their goals will figure out who needs to be in the
   lead for each of them and will make sure every business-
   change effort takes full advantage of their abilities.

Expectations management: We've written what follows as a
handbook, intended for an executive audience. In the interest
of brevity, and of not filling pages with content that's readily
available from other sources, you'll find quite a few topics we
introduce and provide high-level guidance for, without digging
down to the level of providing detailed instruction manuals.

   Our goal here is to familiarize you with the most important
disciplines and techniques needed to achieve intentional business change, not to provide primers for them all.

---

* Not all, of course. A lot of product innovation, for example, depends on
materials science and other technologies that aren't information technologies. There's already an extensive library's worth of books devoted to
product innovation. We aren't going to add to it here.

There's no such thing as an IT project. It's always about business change and improvement.

It's easy to say. And if technique were all that matters, getting rid of IT projects and replacing them with business-change projects wouldn't be all that complicated.

But we aren't just talking about a change in technique. We're asking everyone who touches information technology to think about it differently.

Changing people's attitudes? That's much more interesting . . . not to mention more challenging . . . than just mapping out a new business process or two and asking everyone to follow them.

1

It's Always the Culture

My job isn't to make you happy. It's to not give you an excuse.
—Anonymous help desk analyst

W*e're starting every chapter with an interlude—a tale of either woe or success that is related to the chapter's subject.*

The interludes aren't intended to be Aesop-style fables, perfectly crafted to make the chapter's points, ending with "the moral of the story is . . ." They're complementary to the subject matter, not duplicative of it.

We collected these interludes from seven sources we know and trust. While we've streamlined them, we've done our best to preserve their essence, and our sources have reviewed and endorsed them. In the Hollywood reality hierarchy, each one is "based on a true story"—definitely more accurate than "inspired by a true story" but fictionalized enough to not be unvarnished history.

Interlude: A Cultural Transformation

Back in the last century, the company where I worked was acquired. We were a well-established firm with a loyal customer base and a hard-

*driving national sales team. The acquiring company wanted to trans-
form our successful product manufacturer/sales company into a company
that would assist its customers in achieving their objectives through a
long-term advisory relationship with their company representative.*

No big deal, just move from a culture built on closing sales on as
many of our products as you could while being careful not to give advice to the customer (it was a highly regulated industry) to one of intentionally consultative services with products available, as
necessary—but a rigid, legal requirement to suggest products only
when the customer had a proven need for them.

The acquiring company made a brilliant choice in hiring a savvy
new leader to run the company. Rather than come in with guns
blazing, giving directions on all the things that needed to change, he
came in and spent time learning the current culture of the company.
He then told us something that startled us. We were not going to be
a rules-driven company. You couldn't possibly write enough new
rules/requirements to bring about the transformation required.

We were, instead, going to be a Vision/Mission/Values–based company. He put together cross-functional teams of the "heroes" in the
company: the top salespeople, the top product development people, the
long-tenured, well-respected sales managers, and so on. He then led
them through a set of exercises to define the Vision/Mission/Values of
the new company that we would become. They were concise, mean-
ingful, and focused on helping our clients succeed. I remember those
statements to this day.

Then, to top it off, he really shocked us. He asked us to operate at
an "almost out of control" pace.

For those of us with conservative, risk-averse personality needs, this
seemed like crazy talk. What if we made a mistake and it cost the company money or damaged the brand?

The new leader made it clear to the entire leadership team that
any "mistakes" would be viewed through the filter of "Could you

demonstrate that what you were doing was consistent with the Vision/Mission/Values and focused on serving customers?" If so, all would be forgiven. Learn from it (don't repeat it) and move on.

It should be noted that there were multiple instances where this guidance was validated. As always, there were also instances where actions driven by a different agenda were taken. Those errors were not forgiven and people were held accountable.

We quickly transformed and created a new culture where staff internalized the company's purpose and held each other accountable. In short order, our division became the crown jewel of the parent corporation. Those of us who participated in this transformation marvel at how much we were able to change as an organization and how much we learned. We have all taken that knowledge and experience with us to subsequent organizations and positions; however, whenever we reconnect through industry or social events, we invariably reflect on what an incredible, rewarding time it was to participate in transforming our organization.

In IT's Precambrian* past, its practitioners held exalted status. They were the high priests of the glass house, and supplicants understood that IT operated at a different intellectual level than the rest of the company.

Then came its Pleistocene† fall from grace: IT professionals stopped being mythic gateways to the wonders of computing and became subservient providers to internal customers.

At first this seemed to be just another management fad, to be chuckled at but not taken too seriously.

But what started as a ridiculed change in vocabulary became, over the span of decades, an ingrained part of the IT and busi-

* A long time ago.
† Still a long time ago, but not as long.

ness management cultures . . . and subcultures . . . in most of the world here in commerce's Holocene epoch.*

As is the case with most business change, culture is less of a barrier or enabler than it is the lead story.

And in spite of what you might think, culture is not an intangible I'll-know-it-when-I-see-it "warm fuzzy." It's an aspect of the organization that's just as susceptible to engineering as its processes and technologies.

Susceptible, but, because those pesky human beings are involved at every step, engineering culture calls for considerably more patience.

What Is This "Culture" Thing of Which You Speak?

Business leaders talk about culture all the time. What do they mean when they use the word? Chances are, they aren't all that sure themselves. So let's get precise.

Culture is how we do things around here. It's how we think about things. It's our attitudes, shared knowledge and values, expectations held in common, and specialized vocabularies. It's the organization's unofficial policy manual—a collection of unwritten rules that are rigidly enforced by the inexorable power of peer pressure.

It is, in operational terms, *the learned behavior employees exhibit in response to their environment,* their environment consisting largely of the behavior of the employees who surround them.

That is, it's the learned behavior employees exhibit in response to the learned behavior employees exhibit in response to the learned behavior . . .

---

* Oh, c'mon. Google it.

FIGURE 2 Culture is the learned behavior people exhibit in response to the learned behavior people exhibit in response to the learned behavior people exhibit . . .

It's a self-reinforcing feedback loop, which makes it highly stable and difficult to change—frustrating when you want change to happen, but welcome when it lines up with what you're trying to accomplish (see figure 2).

How to Describe Culture

Before we can talk about how to change culture, we first have to know how to describe it. We need a way to describe the culture we have, the culture we want, and the difference between the two.

Go back to the operational definition—the learned behavior employees exhibit in response to their environment. *Environment* in this context consists of situations. *Behavior* is how employees respond to situations.

The way to describe culture is as a set of situation/response statements.

But there's a caveat to this: it's easy to fall into the trap of thinking you're describing culture when all you're doing is writing procedures.

Let's start by trying to describe your help desk's culture:

> *Situation:* The automated call distributor (the system call centers use to queue up calls) routes a caller to a help desk analyst.
>
> *Response:* The help desk analyst (1) reads the greeting script, (2) records the caller's name and employee ID, (3) looks for open tickets associated with the employee ID, (4) and so on.

This is entirely unhelpful. It describes procedure, not culture. Here's an entry that helps describe help desk culture:

> *Situation:* An employee describes a problem whose solution is, to the help desk analyst, simple and obvious.
>
> *Response:* The analyst rolls his eyes and, while patiently explaining the solution to the caller, makes a mental note to share the call with his lunch buddies as another dumb-user story.

Descriptions of culture are behavioral, but they reflect attitude.

The Culture We Have; the Culture We Want

When a company describes its major projects in terms of IT delivery, a number of mental habits are commonly in play. Here are a few samples, contrasted with the culture in play in companies that describe them in terms of their intended business changes. Your mileage may vary. If it does, we encourage you to develop equivalents that fit your actual situation:

- Where there are IT projects, **business executives** view project proposals with suspicion. They see their job as screening out bad ideas, which is why they insist that the CIO provide a hard-dollar return on investment, typically calculated as the number of warm bodies to be laid off multiplied by their annual compensation plus benefits.

  Where there are no IT projects, business executives view project proposals as opportunities to improve how their company conducts its business. They see their job as helping good ideas succeed. Because of this they insist that all project proposals be described in terms of business change, described that way by a business sponsor who is enthusiastic about the possibilities and supported by an IT SME who attests to the project's technical feasibility.

- With IT projects, **business managers** are busy people. They look for parts of their operation that should be automated, figure it's up to IT to figure out how to provide the automation, but don't have much time to spend with IT—they want IT to just tell them when it's done.

  With no IT projects, business managers are just as busy. But they make time to talk with IT's internal business consultants, asking them for ideas on how to make their areas of responsibility more effective. And, they rearrange their calendars so they can stay involved throughout the process of making the ideas real.

- In IT project-land, a **business analyst** wouldn't think of visiting, say, the company's raw materials warehouse when designing a warehouse management system. A business analyst's responsibilities begin and end with interviews with SMEs to document requirements.*

In the land of truth and righteousness—as you'll read in the next chapter—there are no business analysts. Instead, IT has a staff of internal business consultants. These folks were born with the curiosity gene, which drives them to want to understand in depth the parts of the business they're responsible for, so they can envision a variety of ways to improve things.

- If business change is a battle to be won, **project managers** are the sergeants who win it. That's when there are no IT projects. When there are, project managers are the sergeants who make sure IT delivers its defined work products, and then insist there's nothing left to be done.

- And with IT projects, **IT developers**, as described in the introduction, see their job as translating written requirements and specifications into a bug-free software product. Understanding the business isn't part of their job description; helping change it is Someone Else's Problem entirely.

When projects are defined in terms of successful, intentional business change, in contrast, IT developers spend a lot of time talking with business users because they see no way to succeed without collaborating with their colleagues who work in the affected part of the company.

---

* One of your loyal authors was responsible for designing a replacement for a failed system a business analyst who thought like this had designed. The analyst explained that the warehouse staff had never used the system the way it was supposed to be used. When asked what he'd seen in the warehouse that led to that conclusion, he answered, "Visit the warehouse? Why would I want to do that?" This really happened.

In even more enlightened companies they sometimes become business users for a while so they can envision firsthand what software can do to improve the situation.

- And finally, with IT projects, the projects needed to keep the IT infrastructure current and in good shape are IT's problem, and that includes funding them, on the grounds that if the project doesn't directly benefit my part of the business, I don't know why I should have to pay for it.

After the no-IT-projects culture change, all business leaders will recognize that an obsolete or poorly engineered IT infrastructure is a serious business risk, whether or not it's able to run everyone's applications right now.

You might have noticed that the "with IT projects" examples—those describing the culture we're coming from—all sound pretty bad. That's premeditated.

For the most part, cultural traits are neither good nor bad. They are, however, either consistent with and supportive of the business change you're trying to achieve or run counter to it.

In case the point isn't clear . . . in case you think cultural traits that run counter to what you're trying to accomplish are bad . . . consider the case of World War II. In World War II we considered those resisting change to be the good guys, which is why we called them "The Resistance." Our words for those promoting change were considerably less flattering.

That's the point: while in absolute terms cultural traits are neutral, you're free to choose the language you use when describing both the "from" culture and the "to" culture to emphasize the latter's desirability.

For example, if a CIO was trying to instill an "internal customer service" culture within IT, business manager culture might have been much as described in the *from* version above, but with different word choice: **Business managers** *are busy people. They look for parts of their operation that should be automated, and trust us to figure out how to provide the automation. They don't want us to waste their time. They just want us to tell them when it's done.*

Part of changing culture is choosing how to describe what you want, and what you don't want.

How to Change Culture

Culture, you'll recall, is how employees respond to their environment, most of which is composed of the behavior of the employees surrounding them. Faced with this circularity you might figure culture is either immutable or impervious to deliberate efforts to change it.

It's neither, because when it comes to defining culture, not all employees are equally influential. Put simply, employees in leadership roles have an outsize impact. In addition to those with management titles, this includes employees who have, for one reason or another, taken on aspects of leadership regardless of their official role.

So if the culture isn't what you want it to be, the starting point for your culture change program is a mirror. Look into it and ask the person looking back at you what it is about your behavior that makes the culture what it is.

Understand that and you'll have a pretty good handle on how you need to behave differently to effect the culture change you want.

Mostly, it means recognizing situations in which your response inadvertently models the responses you don't want from the men and women who report to you.*

What it doesn't mean is writing sermons . . . lectures about how everyone has to change because this is the new culture we're trying to achieve. Lecturing people rarely persuades them to change their misguided ways. When you lecture them you absolve yourself of responsibility for a culture that is, in very real terms, something you and your fellow leaders created through your responses to the situations the organization faces on a regular basis.

This doesn't mean culture change should be covert. Far from it, you want everyone to be clear about what you're looking for. Word choice counts here, so choose your vocabulary carefully. A not-directly-relevant example: Your dictionary probably lists "happy," "glad," and "cheerful" as synonyms. They create very different impressions, though, and these impressions matter a great deal when describing the culture you want.

Keeping this in mind, here's a starting point for a glossary to support the change in your company's management culture, *from* one built around the IT organization delivering information technology to its internal customers through the mechanism of

---

* Please note: not the "resources" who report to you; not the "rank and file," "great unwashed," "grunts," or some other disrespectful or dehumanizing characterization. Among the many aspects of the culture change you're going to want is establishing a culture of adulthood. Achieving that depends a great deal on treating everyone who works in your business as an adult human being.

They're men and women. Start referring to them as such. You'll be surprised at the impact it has.

IT projects *to* one in which IT collaborates with everyone else to achieve intentional business change:

- **Customers** are people outside the company who make buying decisions about the company's products and services.
- **Colleagues** are coworkers throughout the company who collaborate with each other to win customers, retain them, and increase the company's share of their wallets.
- **Collaboration** is the preferred style of interaction among colleagues, not negotiation. Not even a win-win negotiation outcome, because win-win still means you're still on opposite sides of a negotiating table.
- **Processes** are workflows optimized for repeatability, scalability, adherence to specifications, and overall throughput. They're recipes and assembly lines, where following the right steps in the right order yields the desired results. In general, information technology drives and orchestrates business processes.
- **Practices** are ways of working that are less rigidly specified than processes. They are optimized for flexibility and rely on employees' skills, knowledge, and good judgment to yield the desired results. In general, when it comes to business practices, information technology provides a toolkit whose value lies in helping practitioners be more effective.
- In the end, most **business change** comes from improvements to business processes and practices. These improvements should be focused on improving the value provided by the company's products and services to its customers, its customers' perceptions about and recognition of that value, and the desirability of buying from your company rather than from any of your competitors.

To Sum Up

Culture is "how we do things around here." It's the shared unconscious thought processes that reinforce the status quo, which is why culture change is essential for the *status futurae*—the change from IT projects to those that achieve intentional business change. It's the shift from some essential element of a change being Someone Else's Problem to all elements of business change being everyone's problem.

Culture is rarely good or bad. Mostly it's a good or bad fit to the current situation and to your view of what you want the future situation to look like.

Most important of all, without exception business culture is the result of how leaders behave.

Remember these essential points and you'll start off headed in a positive direction. Because for companies trying to escape the IT projects trap, culture isn't a barrier or enabler.

It's the lead story.

If You Remember Nothing Else . . .
- Culture changes by example, not by proclamation. As a leader, your most important culture-change tool is a mirror.
- Culture is "how we do things around here." For a business to be successful at intentional business change, a shared focus on the desired business change has to be how we do things around here.
- The idea that IT should be like a business supplying its products and services to internal customers is deeply embedded in the IT and business cultures of most companies. It just might be the biggest barrier to overcome on the journey to excellence at achieving intentional business change.

What You Can Do Right Now
- Publicly retitle all projects whose name is based on IT delivery, and explain why.
- Stop calling anyone a "customer" who doesn't make or strongly influence buying decisions regarding the company's products and services.
- Monitor your own behavior and conversations for accidental encouragement of the IT projects mind-set.

2

The New Business/
IT Conversation

God is in the details.
—Gustave Flaubert

The Devil is in the details.
—Friedrich Wilhelm Nietzsche

What are God and the Devil doing in there—
arguing theology over a beer?
—Bob Lewis

Interlude: This Isn't New

Back before methodology was a word, I found myself in front of an IT project of some size and scope. My job: design the new system, as in document the requirements and create the specifications.

I was, I can now admit, completely unqualified for the job. Compounding the felony, in those days of my youth I was utterly unwilling to admit my ignorance about this or, for that matter, any other topic.

Which was why I didn't do what any sensible programmer would have done, which would have been to head straight to the library. Instead I asked myself what I was qualified to do that would seem plausible enough that I could fake my way through the assignment.

The answer, as I explained to my possibly too-trusting manager, lay in the anthropology classes I'd taken in college. These provided a framework for interviewing respondents, documenting and organizing their knowledge.

And so, instead of asking what we now call business SMEs the questions other business analysts asked when interviewing business users before designing their systems, I asked them to explain what they did every day to get their jobs done. In follow-up conversations we discussed what software could automate and what it couldn't, and what information the software would have to manage in order to automate it.

Eventually I wrote it all down—a description of how to handle purchasing, issuing, and receiving for the company's sixteen "nonstrategic" inventories and how the new system fit in.

Through a combination of dumb luck and a highly supportive business sponsor (we didn't know that's what the role was called, but that's what he was) we built the system and implemented the business processes it supported. Remarkably, it all worked as planned—not just the software but the whole new way of handling purchasing, receiving, issuing from stock, and overall inventory management.

The transformation from chartering IT projects to achieving intentional business change starts with a simple-sounding adjustment: Instead of asking a so-called business customer, "What do you want the system to do?" everyone in IT will get in the habit of asking, "How do you want your part of the business to run differently and better?"

Depending on what happens next, the IT representative might add, "I have some tools that might help you figure that out."

That's where the simplicity ends and a deep conversation about business function optimization, experience engineering,* or decision support begins.

These are the three major types of business change that can benefit from better information technology. **Business function optimization** is about how the work of the business gets done and can get done better. **Experience engineering** is self-defining—it's about improving the experience everyone has when doing the work of the business.† **Decision support** helps decision-makers make better decisions by making data and data analysis more available, reliable, and useful.‡

Collaborating with business executives and managers to design each of these in achievable terms will become the basis of the new business/IT conversation and the standard of competence for the new business analyst, who will often be retitled "internal business consultant."

Making them happen will become the standard of competence for the IT organization.

One at a time . . .

---

* This is often called "customer experience engineering," but as with IT leadership's oft-given message that everyone in the department should have a "customer service attitude," the word "customer" adds nothing of any consequence to the message while reinforcing a relationship model that should, as already emphasized, be discarded.

Companies generally benefit when IT staff have a service attitude. Likewise, experiences should be engineered.

† When real, paying customers interact with your website or mobile app, they are, in effect, doing some of your company's work.

‡ We'll cover a fourth type of business change (strategic, transformational change) in chapter 3.

Business Function Optimization

Current State of the Art

You're probably more accustomed to the term "business process reengineering." The reason you won't see this term here is that "process" is one of those words that have more than one meaning, depending on who's using it and how precise they want to be.

In this book we use it to mean a series of planned steps that lead to consistent, repeatable, predictable results. The gold standard for process-ness is the manufacturing assembly line, which is where most process optimization theory comes from.

Modern process optimization practice has coalesced into four major disciplines. They are, in our utterly unbiased perspective:

- **Lean:** reduces waste, and therefore costs; also, incidentally, reduces defects
- **Six Sigma:** reduces variability, and therefore defects; also, incidentally, reduces costs
- **Theory of Constraints:** removes process bottlenecks, improving capacity; also known as throughput
- **Reengineering:** removes large sums from the corporate coffers while increasing risk

Whatever else you do, don't choose one of these as the foundation for your business function optimization program. That seems to be the standard pattern, and it turns these disciplines into competing religions instead of complementary problem-solving techniques. The frequent result: an octagonal peg that fits the octagonal holes quite nicely but dents and mars the rhomboidal ones.

In case the point isn't clear: imagine your business processes are, as is the case in most businesses, good enough to get the job done but not so good that significant improvement isn't possible. Do you really think the problem with all of them is excessive waste? Or a too-high defect rate? Or insufficient capacity?

In our experience, and, we think, in yours, different business functions, which have different purposes, will have different flaws that require different techniques to address them.

Once you understand this, you'll understand the next challenge you'd face if you were to adopt one of these disciplines as your corporate business function improvement religion: the expertise needed to establish these programs doesn't come cheap, even if you choose only one. Develop the expertise you need so as to apply the right one to the right problem and you'll find you've invested quite a bit.

Which will then bring you to the next hurdle: for each business function you'll find yourself with each process improvement discipline's leaders insisting theirs is the right one for the job.

Another reason not to start with these: they're all *process* improvement disciplines, whose practitioners rarely acknowledge that not all business functions are best implemented as processes in the first place.

And one more: none of these disciplines provides tools designed to leverage the capabilities that new information technology brings to the business.

Business Function Optimization: Starting the Conversation

We've found that one of the best ways to approach the subject of business function optimization is to start by explaining the difference between business *processes* and business *practices*.

Business processes are, as noted above, assembly lines. Perform each step correctly and in the right sequence and you can't avoid achieving success.

Business practices, in contrast, depend more on knowledge, experience, and judgment than on following a fixed set of steps. Project management is an excellent example. All good project managers know the essential steps they need to follow in order to bring a project to successful conclusion. They also know that following the right steps in the right order is just the ante that gets them into the game.

Unlike those who execute the steps in a business process, who require training in the steps they execute, in a business practice the practitioners require quite a lot more: street smarts to accompany their book smarts, judgment, and the bag o' tricks that comes only from experience.

Business practices rarely follow a sequence of simple steps. The sequential steps are broadly stated categories of action (gather information), not specific actions (insert flange A into slot B). It's up to the practitioner to figure out exactly what information must be gathered at this time, and what the best ways of gathering it are.

One more point: the question of process versus practice isn't a matter of one-or-the-other categorization. They're the poles of a continuum of possibilities.

Extending the Conversation: The Six Dimensions of Business Function Optimization

Message from Bob and Dave: if you ignore the entire rest of the book, don't ignore what follows.

When it comes to process management, a very old wisecrack has it that you can make things cheaper, faster, or better— pick two.

The concept is right. When you try to optimize a business function you face trade-offs among the different parameters you could optimize for.

But to make the concept useful you need to refine it a bit, because each of these three business function characteristics has two separate and independent aspects to it. You can, in fact, optimize a business function for any of six different dimensions:

- **Fixed cost:** the cost of turning on the lights; the onetime investment in infrastructure you need to make for the process to work as you want it to work.
- **Incremental cost:** also known as marginal cost, the additional cost needed to output one more of the work products the business function exists to produce.
- **Cycle time:** the total time that elapses between starting to work on a product and it rolling off the factory floor.
- **Throughput:*** the number of products that roll off the factory floor in a unit of time.
- **Quality:** in plain English, a highly ambiguous word. Here we use Philip Crosby's definition: quality means adherence to specifications and, from the perspective of business function optimization, addressing any elements of the business function that cause defects.[1] This differentiates quality from . . .
- **Excellence:** in plain English, another highly ambiguous word, often used as a synonym for quality. Here,

---

* Here, we consider throughput and capacity to be synonymous. They aren't actually quite the same thing—throughput is the actual number of work products the business function outputs in a unit of time, while capacity is the maximum potential throughput. But as throughput can be measured while capacity can only be extrapolated, for the purposes of this conversation, optimizing for one or the other is pretty much the same.

excellence means the presence of features and characteristics customers value in the business function's work
products and the business function's adaptability—the
extent to which it can adjust, tailor, and customize as
circumstances demand.

What's most important in conversations about the six dimensions of process optimization is ranking them. That's because
the choices available for improving any one of them inevitably
results in trade-offs among the others.

Imagine, for example, that quality is, for a given process, your
top priority. One of the most popular strategies for maximizing quality is to reduce excellence—to prohibit tailoring and
customization and turn down most requests for exceptions.
Quality would outrank excellence.

And if you consider excellence to be your second-highest priority? That probably means increasing physical inspections—a
step that results in higher incremental costs.

Or, imagine that for a different process, minimizing incremental cost is most important. This almost always requires
investments in systems and other infrastructure—reductions
in incremental cost call for increases in fixed costs.

And so on.

Incrementalism versus Starting Over

When collaborating with front-line business managers—the
folks directly responsible for optimizing business functions—
an important dimension of the conversation should be about
incremental optimization versus starting over.

As a general rule, incrementalism should be the general rule.
It entails less risk, delivers business results faster, and is far less
disruptive besides.

Starting over is the right choice in just a few situations: (1) for strategic reasons the company has decided to replace a major business system, which means the system on which the business function runs won't be there anymore for it to run on; (2) the business process in question follows pretzel logic that's a response to major system deficiencies; and (3) after multiple mergers and acquisitions, the company has ended up with multiple versions of a business function, has now decided to centralize or standardize, and for political reasons or legitimate dissatisfaction with all versions has decided to redesign the business function from scratch.

One more, and it's a tough one to spot: (4) the way you wash your clothes right now is to beat them against rocks in the river. Incrementalism would most likely lead you to program robots to beat them against the rocks more efficiently. It's hard to see how incrementalism could lead you to invent a washing machine.[2]

Incrementalism: Theory of Constraints Revisited   Eliyahu Goldratt's Theory of Constraints[3] is usually thought of as a way to improve throughput. It works by identifying the worst process bottleneck,* fixing it, and then identifying the worst remaining process bottleneck.

That's about right, except that we need to generalize "bottleneck" so it isn't limited to ways to improve throughput.

Here's our modified take on how to go about it:

Step 1: Rank the six dimensions of optimization in descending order of importance.

---

* We prefer *bottleneck* to Goldratt's *constraint* because in other contexts we differentiate between problems and constraints, the difference being that we can solve problems. In this usage, constraints are conditions beyond our control that we have to work with or around.

Step 2: Decide whether any or all of the top three dimensions are unsatisfactory. If none of them is unsatisfactory, be happy and find something else to occupy your attention. Otherwise, any highly ranked and unsatisfactory dimension is called a *pain point*.*

Step 3: Map out the business process. We've found that a combination of black-box analysis, which describes processes in terms of their inputs and outputs only, and so-called swim-lane diagrams† for describing the actual process flow is a good way to go about this. In any event, process mapping is a vital skill for any business analyst who wants to become an internal (or, for that matter, external) business consultant.

Beyond this quick sketch, techniques for mapping business processes exceed the scope of this book, as they're described in exquisite detail in many other existing works.

Step 4: Identify the worst bottleneck steps in the process map, with bottleneck defined as a process step that causes a pain point.‡

Step 5: Fix one of the worst bottlenecks. If you can't fix a bottleneck without changing or replacing one or more

---

* In this book. Most consultants call anything anyone gripes about a "pain point," which might be why so many business improvement efforts base their priorities on whatever sticks to the wall instead of sound engineering.
† The correct name for these is "Rummler-Brache diagrams," to give credit where it's due. If you don't know what these are under either name, they're a handy tool for describing work as it flows from one step to the next and from one actor to the next.
‡ That thought you were holding? Goldratt's original Theory of Constraints assumes the goal of all process optimization is improving throughput. We're applying the same pattern to a wider range of optimization possibilities.

business applications, as is the case more often than not, work with IT to change or replace them.

Step 6: Loop to step 4 until you reach the point of diminishing returns.

With incremental business function optimization, information technology changes are a consequence of desired business changes.

Starting Over: Business Function Replacement, Version 1— Designing and Building Everything from Scratch   This isn't a treatise on business function design. If you've decided to start over and redesign from scratch—to reengineer—here's a quick sketch:

Step 1: Black box. Create an input/output view of the function, or, more accurately, an output/input view. Start with outputs, not inputs, as it's the outputs that are the point of the function.

Step 2: Optimization. Rank the six dimensions of process optimization.

   With these two steps you'll have defined what you're trying to accomplish: the *what*.

Step 3: White box. Use swim-lane diagrams to design the *how*. Four tips here: (1) each swim lane should have between five and nine boxes in it; (2) resist the temptation to add more, instead creating new swim-lane diagrams that drill down into a box in the primary swim-lane diagram; (3) to account for your information technology requirements, add one or more lanes that treat each application as a robot that's just another actor in the process you're designing; and (4) try, starting with the process flow as you want customers or business users to experience it. That's often an excellent way to create the first, top-level white-box description.

Step 3a: Process bypass process. Your goal is to improve things, not to turn your company into a stifling, choking bureaucracy that's perfectly designed to drive customers away in frustration.

Give employees a way to escape from the function's step-by-step design when the design doesn't fit the situation. These are called exceptions, and they happen all the time.

Step 4: The exalted state of good enough. You aren't going to achieve perfection, so don't bother trying. Implement the new function when it, including its process bypass process, is good enough to get the job done. After that, the same incremental optimization method we just described takes over.

Starting Over Version 2: Systems Replacements   Companies the world over are finding themselves trapped by their legacy systems. "Legacy" is, by the way, a strange term to have made its way into our IT vocabularies. In any other context a legacy is something you're delighted to be the beneficiary of. A legacy system, on the other hand, is a leaky boat, becalmed in the Sargasso Sea of your enterprise. It's something you neither value nor can easily escape.

And oh, by the way, one or more critically important business processes rely on it.

Eventually, everyone involved agrees it's time to retire it.

Mistake #1: Looking in the wrong direction. This juncture is one of two places where most companies get it wrong.

They not only define an IT project ("Replace the main-frame" or something along those lines) but define it in terms of where they're coming from, not where they're going.

They aren't even implementing an ERP (enterprise resource planning), or CRM (customer relationship management), or warehouse management system. They're bent on retiring the mainframe. And so they do, eventually. When they do, they "modernize" the system, which usually entails replacing tens or hundreds of thousands of lines of batch COBOL code with tens or hundreds of thousands of lines of batch Java or C# code, proudly deployed in "the cloud," as if that adds any business benefit.

It doesn't. The sole business benefit is a modest reduction in software license fees, with little or no additional business function optimization; not even much in the way of additional future flexibility.

Mistake #2: Asking the wrong question. Imagine your company avoids mistake #1 and decides what it's moving to. For convenience, imagine the plan is to replace its legacy systems with a modern COTS (commercial off-the-shelf software)* or SaaS (software as a service) ERP solution.

That's when companies typically face the question of whether to implement the new system "plain vanilla" or with "chocolate sprinkles"—in non-gelato terms, whether to configure the application to support the company's current business functions or to force every business

---

* Yes, it should be COTSS. But it isn't, and there's nothing any of us can do about it.

function manager in the company to adapt to the new system's default way of doing things.

That's such a wrong question that many businesses never recover from asking it.

It's the wrong question because it ignores the six dimensions of business function optimization.

As a wise IT master once explained, software is just an opinion. It's an opinion about how your business should handle whatever process or practice the software is designed to support.

Which leaves you to answer this question: Is the software's "opinion" better or worse than your company's current opinion about the subject?

If it's better, your organization should unhesitatingly "change its mind," which is to say, it should adopt the process or practice embedded in the software. If it's worse, you should reconfigure the new software to support how you do things right now.

Which leaves the question of how to go about comparing the two. It's a question that's easily answered, at least in principle.

Start with the same six dimensions of optimization we've been beating you about the head and shoulders with the last few pages, and for the business function in question, rank them in order of importance.

Next, measure how they perform with respect to the dimensions you've ranked most important. That's your baseline—data your business collaborators presumably collect as a matter of course.

They don't? Really?

Oddly, we've found that relatively few organizations do a good job of this. In any event, do what you have to

so you know how the current business function performs.

Then . . . and this is the hard part . . . find a way to
simulate your new application's embedded approach to
doing the same work, and figure out how it performs
with respect to the dimensions of optimization that are
most important for you.

Implement whichever approach will perform better.

The decision isn't plain vanilla versus chocolate sprinkles. Call it mint chocolate chip versus mango sorbet.

We do, by the way, recognize that persuading everyone won't be as simple as just showing everyone how the
numbers compare. It should be, and if your organization
has a culture of honest inquiry (see below), it will be
easier; but people, including you, bring their back-
grounds, experience, and biases to every one of these
decisions.

But as they'll bring them whether you're arguing over
the chocolate sprinkle count or our *deux sorbets* alternative, you might as well go with the sorbets.

Configuration versus Customization

You might have noticed we didn't talk about customizing the
new application as the alternative to plain vanilla implementations. We used the term "configure." The distinction is huge.

Configuration means using tools built into the application
you've licensed specifically to modify its functioning to your
organization's needs, without violating or changing its design
or code in any way.

Customization means fiddling with the code, or perhaps
writing a whole new satellite application that doesn't make

correct use of the integration points included in the package for this purpose when it's needed.

The difference: when the time comes for IT to update the package to a new version, configurations rarely cause challenges. Customizations, on the other hand, greatly increase the cost and risk when updating to new versions.

So while vanilla versus chocolate sprinkles is a conversation to be avoided, customizing a package so as to satisfy a "requirement" is something to avoid except for the most dire of circumstances.

Experience Engineering

Experience engineering is the second type of business change the new IT will support. To a certain extent it's a matter of art and aesthetics, but mostly, done right, it's data-driven engineering. There's a short version and a long version.

The short version: Find a bunch of naturally irritable people. Give them standard tasks to accomplish. Ask them what they find irritating in accomplishing them.

Fix what they tell you, if you can.

That's experience engineering in a nutshell.

It isn't, of course, quite that simple, which leads to the long version.

Experience engineering starts by understanding what users are trying to accomplish and the tasks they have to undertake to accomplish it, and finishes by optimizing each touchpoint.

The rule about tasks is straightforward: users, whether external customers or internal staff, should have as few tasks as possible standing between them and the outcome they want.

Touchpoints are a bit more interesting. They're the intersection of tasks and channels. As an example, a customer might want to schedule an appointment with someone in your company. That's a task—one step in accomplishing something or other. Some customers might prefer to schedule appointments using their telephone. That—executing the task of scheduling an appointment on the channel that is the telephone—is a touchpoint. Scheduling an appointment via online chat is a different touchpoint, as is scheduling one on an online calendar or scheduling one using a mobile app.

As a general rule, no matter what the task, users will have more than one channel available, and they won't stick to the same channel for all the tasks they need to undertake: a first task might be executed via a smartphone app, the second might be via online chat, and the third might be through email or a phone call.

Experience engineering is a matter of making it as easy as possible to get something done, by minimizing both the number of tasks needed to get it done and the level of annoyance resulting from each touchpoint.

Step by Step

For each outcome users might want to accomplish:

Step 1: Develop personas. A persona, in case you're unfamiliar with the term, is a way to categorize users into typical groups that experience your company, and then to characterize them by way of a convenient shorthand. "Demanding Dan" might be one persona, consisting of all difficult but highly profitable customers; "Agreeable Anne" might be the persona for the type of customer who accepts whatever experience they happen to get without complaint, until they get on social media.

These examples are oversimplified. The personas you'll probably want are cross-classifications that include demographics like age, sex, marital status, and income level; psychographics like extroversion/introversion; social media anger management; skill categories of various kinds; or any number of other characteristics.

Step 2: For each persona, consider how they experience your company. Do this in terms of broad outcomes, and consider the tasks they might have to undertake to accomplish each outcome, such as researching products, purchasing an item in person or online, returning an item, or asking for service.

Step 3: Catalog the touchpoints resulting from step 2, each combination of task and channel. When your loyal authors were wee laddies this was simple. Most personas either visited your business or called. Now, depending on the task, any given persona might choose interactive voice response via telephone, calls to your customer service call center, calls to a personal representative, email, online chat, your website, a mobile app, or your company's Facebook page . . . to name some of the more prominent channels.

Step 4: That's a lot of touchpoints. Even more interesting, for a given task, personas will expect to be able to accomplish it regardless of channel, and will expect to perform the next task using whatever channel is most convenient for them, regardless of which channel they used for the previous touchpoint.

This has a significant implication for your company's technical architecture: application capabilities must be made portable across all channels. Otherwise, the cost of supporting them all will increase exponentially as you add channels; meanwhile, the level of user dissatisfaction will increase exponentially if you fail to add them.

Step 5: Don't rely on your own judgment. Don't assume. Verify the validity of your personas by talking to or surveying groups. Validate your touchpoint designs the same way.

Also, don't assume a single set of design principles will be valid for all personas.

For example, a persona who needs to achieve some particular outcome only occasionally might prefer a simple, uncluttered, intuitive user interface when visiting the web page that represents one touchpoint on the path to accomplishing it. A different persona—one who has to accomplish the same outcome a dozen times a day—will almost certainly value the efficiency that comes from having as much information and functionality available on a single page as possible.

Step 6: They don't like you that much and don't want to get to know you better. In the end, the best touchpoint design is to either eliminate the task it supports altogether, consolidate the task with some other task, or at least use what you know to navigate each user to where they're most likely to want to be.

Take, for example, Uber. While you can schedule a pickup for some future date and time, its default assumption is that you want a ride right now—that's where you start.

Also, its experience designers recognized that paying for a ride at the end of the journey is an irritating experience, even with credit card readers now common in the cab's back seat. So Uber consolidated this task with the scheduling task.

It's irritation removal at its finest.

Remember, with few exceptions, your goal, overall and touchpoint by touchpoint, is to irritate as little as possible the customers and users who correspond to each persona. And there are exceptions to the touch-them-least-is-touching-them-best guideline. Destination retail outlets like the Apple Store and Cabela's sporting goods stores are examples; so are theme parks.

But for the most part, irritating users as little as possible is a lofty enough goal.

More Thoughts and Suggestions

Start with customers. Real, paying customers. Why would they want to take any of their valuable time to let you watch them as they try to accomplish standard customer tasks?

Some retailers have set up shopping laboratories—mock retail environments equipped with cameras and sophisticated tracking technologies. They pay customers to shop for stuff in them, tracking their movements, what their eyes fixate on, what kind of signage leads them to put something in their shopping cart, whether they're more likely to buy the same merchandise if it's on an end-cap, and so on.

But no matter how sophisticated the lab, there's no certainty that customers who are willing to shop in exchange for money behave the same way as shoppers who want to buy stuff in real stores.

There's also no certainty that shoppers in, say, a Cleveland suburb (if that's where the lab is located) behave the same way as shoppers in downtown New York, Tallahassee, or Stevens Point.

Hint: Read Paco Underhill's *Why We Buy.*[4] Underhill suggests visiting real stores, finding a quiet spot, and watching real shoppers try to cope with what they encounter.

This, or the equivalent, is good advice for engineering all experiences your customers have with your company, to the extent it's practical, at least.

For example, **when a customer visits your website** (and likewise your mobile app), you can track where they click, how much time they spend on different pages, and what they were looking at when they decided not to buy your merchandise.

Also, when the time comes to redesign your website or mobile app, recruit some friendly customers who are willing to look at your new designs and let you know what they like and don't like about them, in exchange for a modest sum. Or, especially for direct-to-consumer websites and mobile apps, invest more to learn more. Labs where real users are monitored to see how they interact with your website or app have become the gold standard.

When they phone your call center, you can do more than just "record their calls for quality purposes." For starters, you can figure out how often callers make the wrong choice when navigating your menus as well as how often they hit 0 or say "representative" when an automated alternative would have helped them accomplish what they wanted to accomplish. That's along with the standard fare: queue time, abandon rates, and so on.

Beyond all this, artificial intelligence (AI) is encroaching, in the form of chat bots and email autoresponders. These are attractive from the perspective of cost minimization. But remember your goal of minimizing touchpoint-induced irritation. So especially with these AI-based solutions, test, test, and test some more to make sure they aren't noticeably more irritating than their human equivalents.

Speaking of AI, here's a look ahead, and a hint.

The look ahead: the technology is just about ripe for customers telling your automated systems, in plain language, why they're calling.*

You don't want your competitors to get ahead of you with this technology. If you decide you can't afford it, figure out how to train humans to accomplish the same thing. If you can't afford humans either, figure out how many customers you'll lose to competitors that offer this level of service and the cost of that lost revenue.

The hint: never mind the quality assurance recordings. Ask the human beings who staff your call centers what the people they talk to find most aggravating about your company. They know. They deal with your crabbiest customers all the time and they'd be happy to tell you what's making your customers crabby, all without a dime of investment in big-data social media analytics.

And a tip: ask them to track issues as they answer calls throughout the day. The loudest callers make the biggest impression. A small bit of tracking will help everyone keep loud callers in perspective.

And yes, in case you're wondering, we do expect your average business analyst of the future to help guide their business counterparts through this thought process.

Decision Support

And so we get to decision support, the third type of IT-supported business change.

---

* Yes, yes, yes, if you want to be precise, the technology for letting your customers tell you why they're calling has been around for a century. What's new is that the automated systems will soon be able to accurately interpret what your customers say.

"Decision support" is an old but still useful term. IT has been trying for decades to build systems that help executives and managers make better decisions, with limited success at best.

The technology has improved over this span of time, from custom reports written by IT programmers, reading data maintained by the company's business applications; to "user friendly"* report-writers that go after the same data so IT no longer had to be involved; to carefully designed data warehouses and the user-friendly business intelligence tools designed to analyze the contents of the data warehouses so as to make the business more intelligent; to modern big-data repositories contained in "data lakes" that require data scientists to look for useful patterns while making sure the statistical analyses run against what's stored in the data lakes and conclusions drawn from them are valid.

All of which constitutes technological progress. Whole books, of significant size and heft, have been written about business intelligence and the associated IT engineering. Unlike business function optimization and experience engineering, business intelligence implementations for the most part have been about business change from the day they were first envisioned. We have nothing new to offer on that front.

Except for this: none of it is worth a thing without a culture of honest inquiry.

A Culture of Honest Inquiry and How to Get One

Honest inquiry is a matter of embracing the conclusions that result from what the best evidence and soundest logic tell you.

---

* With apologies to *Rocky and Bullwinkle,* to be more accurate, less user "fiendly."

It's a matter of understanding that your gut is for digesting food—your brain is where thinking takes place.

Reliable evidence, and relying on evidence, is vital to making smart decisions in business. In *Good to Great*, Jim Collins quotes Lyle Everingham, CEO of Kroger during its transition from muddling through to twenty-five years of outstanding performance: "Once we looked at the facts, there was really no question about what we had to do."[5] A&P, its lackluster competitor, pretended, creating a new store concept, the Golden Key, to, supposedly, test ideas. Its executives didn't like what the evidence told them, so they closed the Golden Key business.

Kroger had, and apparently still has, a culture of honest inquiry, where executives, managers, and employees do their best to use trustworthy evidence to drive decision-making. Creating a culture like this takes work, persistence, and sometimes political dexterity. Here are some specific measures you can take to foster a culture of honest inquiry in your workplace.

It starts with wanting to know what's really going on out there. Enron and WorldCom happened, in part, because their executives were so busy trying to make their companies look good that they obscured what was really going on to themselves. Your dashboards, financial reports, and other forms of organizational listening are to make you smarter. If that isn't what you plan to use them for, don't bother.

Confidence comes from doubt. Certainty, in contrast, comes from arrogance. If an employee is confident and can explain why, wonderful. If that employee's certainty preempts everyone else's ability to make their case, the employee is on the wrong side of things.

And yes, we are including the company's top executives as employees as we say this.

Start every decision by creating a decision process. You don't have to be in charge to encourage this habit. Just ask the question, How will we make this decision? That changes the discussion from who wins to how to create confidence in the outcome. The results: a better decision, a stronger consensus, and a few more employees who see the benefit of honest inquiry.

Don't create disincentives for honesty. If you ask for honest data and use it to "hold people accountable," you won't get honest data. Why would you? The superior alternative is to employ people who take responsibility without external enforcement, so you don't have to hold them accountable, and to make sure employees who give you honest evidence aren't shot as unwelcome messengers. This works much better and takes less effort.

The "view from 50,000 feet" is for illustration, not persuasion. A high-level strategic view is essential for focusing the efforts of the organization. High-level logic, in contrast, is oxymoronic: detailed evidence and analysis is what determines whether the high-level view makes sense or just looks good in the PowerPoint.

Evidence too far removed from the original source is suspect. Don't trust summaries of summaries of summaries, especially if they tell you what you want to hear. Even with the best of intentions the game of telephone is in play. And many of those trying to persuade decision-makers don't have the best of intentions.

Be skeptical of those with a financial stake in the decision. But don't ignore them. A conflict of interest suggests bias but doesn't automatically make someone wrong. Be wary and dig into their evidence, especially if their evidence is a summary of a summary of a summary—even more so if it tells you what you want to hear.* But if you demonstrate to your satisfaction that they've cooked the evidence . . . go ahead and ignore them from now on. They've earned it.

Beware of anecdotes and metaphors. They're useful . . . for illustrating a point or for demonstrating that something is possible. For anything else you need statistically valid evidence. Yes, someone said there are three types of lies.[6] He miscounted; argument by anecdote is far more pernicious than argument by statistics, and argument by metaphor is even worse. Yes, you do have to understand statistics well enough to evaluate the evidence. Sorry. That's part of your toolkit.

Be alert for "solving for the number." This is a popular management pastime that preceded technology. It's achieved increasingly high levels of false precision with the advent of the electronic spreadsheet, and even more with business intelligence software. It refers to the practice of starting with the answer you want and then fiddling with filters, adjusting assumptions, and, for the ultra-sophisticated, applying various different statistical procedures to your data until you get the results you want.

If you work in a business without a culture of honest inquiry you'll need time and patience to build the habit of rationalism. You won't do so by preaching and lecturing about the general principle.

The way to build a culture of honest inquiry is one decision at a time. Especially, you can help build it by finding opportunities to be persuaded by evidence and logic and by making it okay for employees to change their minds.

And in case you're wondering, yes, when it comes to decision support this is part of the new business/IT conversation. But this is a part that doesn't rest with IT's business analysts.

---

* For more on this subject, google "confirmation bias."

It's a tough conversation the CIO has to have in the executive suite.

In Conclusion

In most organizations, CIOs, IT managers, and especially business analysts sincerely want to satisfy their internal customers. This means getting the product right, which in turn means establishing elaborate mechanisms for figuring out what the internal customers want the software to do.

With a bit less sincerity and a bit more cynicism, all parties would know the answer: these so-called internal customers don't want the software to do anything. They want their part of the business to run differently and better; they want real, paying customers to have a great experience interacting with the company whenever, however, and whyever* they're interacting with it.

And, they want to make more informed decisions whenever it's possible for more and better information to help them do so. That's what the new business/IT conversation will be about.

Information technology will often be part of the discussions.

If You Remember Nothing Else . . .
- The new business/IT conversation begins with IT asking, How do you want your part of the business to run differently and better?
- Business processes (your assembly lines) and practices (your knowledge, experience, and secret sauce) are the two poles of the continuum of how to organize the work that has to get done in an organization. Figuring out where on the

---

* Not a word, but it should be.

continuum a specific business function should be placed is the starting point for making sure it's properly designed.

- You can't effectively design a business function until you've defined its outputs and determined what inputs are needed to create those outputs; you also must know how the six dimensions of business function optimization rank.
- Designing the customer/user experience is complicated. Success starts by setting this goal: make their experience as un-irritating as possible.
- There's no point implementing any decision-support technology until the enterprise has begun to institute a culture of honest inquiry. Decision-support systems and practices are valuable only to the extent they reinforce this culture.

What You Can Do Right Now

- Educate the company's business analysts to stop asking their collaborators in the business what they want the software to do, and instead ask them how they want their part of the company to run differently and better.
- Educate business analysts in the fine arts of process design and optimization, and in experience engineering.
- Educate every manager in the company in the six dimensions of optimization. Ban "vanilla versus chocolate sprinkles" debates when implementing commercial software packages, replacing them with six-dimensions-based process selections.
- In the C-suite, introduce the idea of a "culture of honest inquiry" as a prerequisite for implementing better analytics capabilities.

3

Fixing Agile

Consultant (n) A form of management advisor
specializing in fixing what's broken
by breaking what's fixed.
—Sheldon Seymour Jr.

Interlude: AgileShock

Not too many years ago we had to replace the core applications that
ran a large portion of our business. The applications did what we
needed them to do, but they were built on a last-generation platform
that was quickly becoming a no-generation platform.

Some analysis revealed that trying to shoehorn a commercial pack-
age into our business was a nonstarter for the specialized services our
company offers. Rewriting was our best alternative, no matter how
expensive and risky it seemed.

My team was used to operating as a craft shop, which made sense
on the old platform. We had a clean architecture and seasoned pro-
grammers. The combination let us handle as one-programmer en-
hancements what would have turned into small projects in most
companies.

On top of this, our programmers weren't shy about talking things over directly with the business users. I figured we were already halfway to Agile. Scrum seemed to be the most popular version, so I figured, how could I go wrong?

The project nearly failed within a year.

Here's what went wrong.

The techniques needed to rewrite a major application were outside our day-to-day experience, so I brought in some very talented developers who knew the new technology and had extensive experience with Scrum. My plan was for these experts to train my existing staff on the new technology and on Scrum. We recruited product owners, and the project was off and running.

It turned out our hired experts had no interest in learning our company's collaborative business culture. For that matter they had no interest in learning our IT culture either. Scrum was as much a religion for them as a set of techniques, so within a few months my internal staff started to revolt, claiming the outside experts didn't respect their knowledge and expertise. For their part, our hired experts told me my internal staff were behind the times and refused to change.

I had to make a critical decision. It wasn't hard. I liked our culture—business and IT—and if the outside experts thought it needed to change, there was only one possible answer.

So out they went, and we went in search of a way to fit Agile into the business instead of forcing my business and IT management . . . and I'm including myself . . . to adapt to textbook Scrum.

Then, my team did what my team did best: they figured it out. We discovered there's more to Agile than Scrum, and a different version, called Kanban, fit our way of working quite well.

The result: after losing six full months of actual development time, we brought in the project under budget and ahead of schedule.

Not to mention getting terrific reviews from our business collaborators.

Projects are the way organizations make tomorrow different from yesterday. Project management is the practice (not process) of understanding what work has to be done to achieve the project's goals and making sure that it does, in fact, get done when and how it's supposed to get done.

Within the overall realm of projects and project management, IT projects in particular have acquired a reputation for delivering disappointing results. And those are the more successful ones. Many deliver no results at all.

Which gets us to the subject of this chapter: Agile project management.

There are two types of IT organizations: those that have embraced Agile and those that continue to suffer from a high rate of project failures.

Okay, that's a bit strong, but just a bit. According to the Standish Group, in an analysis of more than ten thousand projects, Agile projects were more than three times more likely to succeed than their Waterfall equivalents (see figure 3).[1]

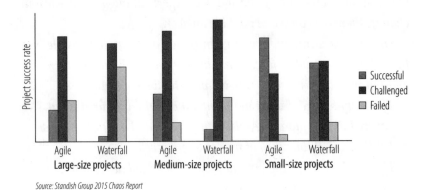

Source: Standish Group 2015 Chaos Report

FIGURE 3 Project success comparison, Agile vs. Waterfall.

Agile is a big honkin' deal. In case you've heard of DevOps*
and are wondering, it's both important and badly misunder-
stood. It's one Agile variant among many, and while not a big
honkin' deal, it's certainly a midsize one.

But it isn't different enough from other Agile variants to
warrant a separate discussion in this book. In this chapter
we focus on the whole family of Agile methodologies, how
and why Agile is superior to the Waterfall methods it's
replacing, and why Agile needs fixing if it's going to survive the transition from IT-as-vendor to IT-as-collaborator-in-
intentional-business-change. Just about everything we'll say
applies to DevOps too.

What Are These Waterfall and Agile Things of Which You Speak?

In case you haven't bumped into Agile or been bumped by it,
it's a way to design and build software that's proven vastly superior to the traditional way of going about it.

The traditional alternative is called Waterfall, because it divides the process into phases; and once you finish a phase . . .
once you go over the Waterfall . . . it's expensive and disruptive
to go back to the top.†

Waterfall methodologies typically divide application development into Feasibility (a silly name, as with rare exceptions
the actual programming is always feasible), followed by
Requirements Gathering, External Design (what it's supposed to

---

* In case you're wondering just what the heck is DevOps anyway, don't
worry about it. From the perspective of achieving intentional business
change, it's just one more variation on an Agile theme.

† No, it isn't called Waterfall because once you go over it you crash onto the
pile of rocks at the bottom and aren't likely to survive the experience.

look like), Internal Design (think block diagrams), Specifications (blueprints), Development, Testing, and Production.

Waterfall techniques aren't limited to application development. Any project that divides work into sequential phases, each of which must be completed before the next one can start, is by definition Waterfall.

Which means what follows matters even if you don't ever expect to have anything to do with implementing information technology (which probably means you're a year or less away from retirement, but who are we to judge?).

So even if you're entirely uninterested in application development, PAY ATTENTION to . . .

Waterfall's Wrong Assumptions

Anyone who has ever planned anything by dividing the work into sequential stages has, knowingly or otherwise, made use of Waterfall techniques.

Waterfall seems so logical. But logical or not, Waterfall techniques rarely work—not because of bad execution but because Waterfall is built on several wrong assumptions, which matter no matter what you're dividing into stages. They are:

- **Constancy:** For Waterfall to work, the assumptions, conclusions, decisions, and designs you made yesterday will have to fit the world as it will be tomorrow. That's fine for short projects, short meaning certainly no more than six months; three is better. Beyond that, as everyone knows,* change is the only constant in business and the rate of change is accelerating.

---

* Usually, what everyone knows is wrong. Not in this case, though.

Whatever decisions a design team makes, many will change from right to not so right long before the time comes to put whatever it is the team is creating to productive use.

With Waterfall, whatever goes into production is designed for the world as it used to be, or maybe for a three-year-old three-year forecast of how the world should have turned out. See "Procrastination is a virtue" below.

- **Envisionability:** Okay, we made the word up. So sue us. Waterfall means you design everything before you build anything. For this to work out, designers have to correctly envision, in detail, how whatever it is that's being implemented should work.

  Good luck with that. Someone a long time ago[2] researched people who designed their own homes. It turned out that most of them designed, built, moved into, and then sold or tore down the first two because they didn't like living in them.

  When envisioning even something as familiar as living in a house takes three tries to get it right, why would anyone expect developing software or implementing a business strategy to be any different?

- **The SSC ratio:** This is the "Stay the Same/Change" ratio.[3] It's how long whatever you're implementing will continue to be relevant and useful—how long you and it will be able to stay the same—compared with the time needed to implement it. Think of it as how long food you grow and prepare lasts before it becomes too moldy to eat.

  The acceleration of change means the time-to-mold metric is getting shorter, but the time needed to plant, grow, harvest, and cook isn't.

Okay, we lied. These assumptions aren't always flawed. Quite the opposite. Consider the case of the successfully constructed office building. You'll find the above assumptions fit construction projects quite nicely.

Which shouldn't be surprising. A lot of project management theory started out by looking at successful construction projects. Then the analogists got hold of things, and, as is usually the case, it all went sideways: "Designing and building software must be just like designing and building a skyscraper, you see, because both of them have to be designed, then built."

If we're going to use construction as our metaphor, designing and building software is more like designing a building for a piece of land and then having to build the building on a different piece of land, in a different climate, with building codes that change halfway through the project.

So no, the one isn't just like the other.

Agile's Essence

Agile methodologies, and there are at least a half dozen of them, don't make these assumptions. At its core, Agile is built on a few key principles that, with apologies to the Agile Manifesto,[4] are, for our purposes at least:

- **Programmers don't need translators.** In chapter 2 we talked about the business analyst of the future. Now let's be honest about the business analyst of the past. Much of their trade was based on the utterly preposterous premise that programmers need someone to listen to business users, translate "UserSpeak" to "ProgrammerSpeak," then listen to programmers' replies, translating their words back into UserSpeak.

We've worked with hundreds of programmers and other technical professionals throughout our careers, and what's amazing about them is how many have friends, spouses, children, and pets who aren't also technical professionals and yet are entirely capable of communicating with them.

In Agile projects, developers talk with businesspeople, and they do so frequently and informally. They also learn the business to the point that they can assist in envisioning solutions that are both creative and realistic, as the developers have their toolbox in mind as they talk.

- **Programmers often do need reorientation.** Programmers can and do have conversations with business managers and end users right "out of the box." That doesn't mean they automagically acquire empathy for what their change collaborators have to deal with every day.

  A technique that works in many situations is to reassign the development team to work in the affected business areas so that they have to do the actual work using the actual processes and systems targeted for replacement. They'll get to know the end users and managers as actual human beings, and they'll experience their frustrations firsthand.

- **Big solutions that work start out as small solutions that work.** Even big legacy systems obeyed this rule. Look at their history and you'll find each one started out as a small system, the result of a business manager asking a programmer, "Can you make the computer do x?" And, once the programmer got the computer to do x, asking, "Now can you get it to do y?" Y built on x, z built on x *and* y, and after a year or three a big system was up and running.

  This worked for two reasons. The first is envisionability: small enhancements are easier to envision than big systems.

The second is a human characteristic: people, including developers, stay motivated when they see progress. Building a big system one small piece at a time means progress is constantly apparent.

Agile turns this into a formal methodology.

- **Procrastination is a virtue.** More precisely, delaying decisions as long as possible, so that you're relying on your ability to predict as little as possible, is a very good idea. If the future you have to be right about when making a decision is one year ahead, you'll base your decision on a more accurate forecast than you would looking three years ahead.

  Don't, however, procrastinate so long that your decision becomes irrelevant. Also, don't procrastinate about building flexibility into the application's core architecture.

  Without this, Agile can result in teams painting themselves into a metaphorical corner. With it, they can change direction quickly as business situations change.

- **Small chunks are less risky than big chunks.** The word is "incrementalism," and in terms of both risk per chunk and bang per buck, incrementalism wins every time.

  It must be said: in proportion, small bangs deliver bigger bangs than big bangs, and deliver them more reliably.

  The challenge is getting the small chunks to come together, jigsaw puzzle fashion, to make a picture that works when viewed as a whole as well as in the details.

  The secret to this is . . .

- **Iteration.** Imagine you're playing miniature golf. The hole presents the usual mini-golf challenges of slopes, obstacles, caroms, and such. Waterfall means hitting balls from the tee box until you get a hole in one. Agile's iteration principle means playing miniature golf as we're accustomed to it. We start in the tee box and then hit each shot from where the previous one ended up until the ball falls into the hole.

At least, that's part of the secret. Agile does have its own challenges IT has to address, most of which have to do with coherence. Coherence challenges enter into Agile projects at two levels.

The first is technical, a matter of making sure all the different pieces are built to a single set of engineering standards. The second is aesthetic, a matter of making sure different developers adhere to a common user interface style. Both are manageable enough that for the purposes of this book we don't need to explore them in more depth.

But if your IT organization hasn't yet taken the Agile plunge, it will need to explore these challenges in more depth.

One other point: if your IT organization hasn't already made the shift from Waterfall to Agile, you'll need that to happen. This isn't a book about accomplishing the IT Waterfall-to-Agile transformation. If you do need to head down that path, keep one point firmly in mind: as with any other shift that entails changes to core and often hidden assumptions, culture change is the lead story, not process, and in fact it's fair to say many of the unsuccessful attempts at this transition failed because they focused on process and not culture.

But

Agile as usually practiced has three . . . well, they aren't fatal flaws, but they're still serious challenges.

Flaw #1: Agile projects are still IT projects. Agile projects deliver a software product to IT's "internal customers." Agile violates the no-such-thing-as-an-IT-project rule.

And it's quite explicit about it. It starts with Agile's insistence that there be a "product owner" in the business who has the authority and expertise to make decisions about product features and priorities.

It ends with how Agile practitioners express the items in their to-do lists . . . as "user stories," which are descriptions of what the software should do from a user's perspective, and why.

It's about the software, not about the business change.

Flaw #2: Agile methodologies address what IT does least. A problem that might appear trivial but that actually matters quite a lot is that most IT shops follow a straightforward principle—they buy when they can and build when they have to.

It's a logical consequence of simple long division. When a company builds an application, it pays the entire cost itself. When it buys an application, it shares that cost (indirectly) with all of the other customers of the company that sells it.

Also, when a company decides to buy, whether it's software installed in the company's data center (COTS) or software that runs in the cloud (SaaS), the design and programming are finished, while when IT builds from scratch the design and programming haven't begun.

As you're buying a finished product, business benefits should show up more quickly than when you build from scratch.

So while there are trade-offs, in most enterprises, most of the time, the buy/build principle provides valid guidance.

And yet, most Agile variants are about building software from scratch, even though what IT does the most is install, configure, and integrate packages. Not only do most Agile practitioners ignore this little fact, but when companies install, configure, and integrate packages, the way they go about it, as described earlier in this volume, often intentionally minimizes the value they get from their effort. They do their best to make their shiny new system behave exactly like the old, tired one it's replacing.

Flaw #3: The strategy-to-action impedance mismatch. Truth be told, few businesses are competent at turning their strategic intent into an actionable plan of any kind. All too often, the so-called strategic plan is little more than a touchstone—a description of the hoped-for future, to which all proposed projects should "be in alignment," which is to say, they have to be consistent with the strategy.

But any manager with a gram of creativity knows how to write a two-paragraph explanation of how whatever it is she wants to do is well aligned with the company's strategy.

And even when a project really is well aligned with strategy, that's very different from advancing the strategy.

Some companies are better than this. They do know how to turn their strategies into plans of action.

But their plans, while not bounded by IT delivery, are still Waterfall plans of action. They define strategic programs, which are composed of major initiatives, each of which consists of a project road map, all nailed down into a linear plan.

Organizing change like this is, we hasten to add, nothing to sneeze at. Building out this sort of plan requires vision, courage, and trust, not to mention quite a lot of cash and a willingness to say no to project proposals that usually are worthwhile, not to mention strategically aligned.

And yet, Waterfall strategy-to-action plans suffer from the same flawed assumptions as most other Waterfall-based projects. Courtesy of the SSC ratio, the methods that made sense when five-year or ten-year strategic plans were useful are no longer sustainable.

But imagine for a moment they were, that you could create a Waterfall-oriented five-year strategic plan and it would hold up for enough years to matter.

Now imagine that, as is the case with most other forms of business change, your business strategy depends on information technology to succeed.

As you've seen, Agile application development is far more reliable than Waterfall application development.

Which leads to a sort of impedance mismatch between how you're organizing and managing strategic change and how you're organizing and managing the application delivery it depends on.

Enter something called the Scaled Agile Framework (SAFe; the mysterious "e" at the end is gratuitous). Its purpose is, as its name suggests, to allow Agile to scale to strategic-program-size efforts. It is, regrettably, complicated where Agile is simple, high-overhead where Agile isn't, and dependent on centralized oversight where Agile is distributed.

It is, to our eyes at least, Waterfall with a thin Agile skin on it.

It's also probably necessary if Agile is to be synced up with strategic program planning and management, at least as it's usually practiced. And yes, we do provide an alternative. Read on.

Fixing Agile

If you've managed to slog through all of the above, you'll see that for Agile to go beyond product delivery to be the methodology of choice for achieving intentional business change, we have some work to do. Not overwhelming work, but work nonetheless. Specifically, we need to adjust Agile to:

- Deliver business change (of course)
- Deliver business change in tandem with COTS/SaaS package implementations
- Sync up with strategic plans

What follows are sketches, not instruction manuals. Our hope is that you and your fellow organizational change drivers will find them useful as guides for developing and sharing your own techniques.

One at a time . . .

Modifying Agile to Design and Deliver Business Change

Good news! We mostly covered this in chapter 2. To save you the effort of skipping back and forth:

Step 1: Rank the six dimensions of optimization—fixed cost, incremental cost, cycle time, throughput, quality, and excellence—in descending order of importance.
Step 2: Decide whether any or all of the top three dimensions are unsatisfactory. If none of them are unsatisfactory, be happy and find something else to occupy your

attention. Otherwise, any highly ranked and unsatisfactory dimension is called a *pain point*.

Step 3: Map out the business process. We've found that a combination of black-box analysis, which describes processes in terms of their outputs and inputs only, and so-called swim-lane diagrams for describing the actual process flow is a good way to go about this. In any event, process mapping is a vital skill for any business analyst who wants to become an internal (or, for that matter, external) business consultant.

Beyond this, as we already explained in chapter 2, this book isn't the place for a detailed account of this well-explored discipline.

Step 4: Identify the worst bottleneck steps in the process map, with bottleneck defined as a process step that causes a pain point.

Step 5: Fix one of the worst bottlenecks. If you can't fix a bottleneck without changing or replacing one or more business applications, as is the case more often than not, work with IT to change or replace them.

Step 6: Loop to step 4 until you reach the point of diminishing returns.

This process is intrinsically Agile: It's iterative and incremental. It augments Agile's Product Owner role to the status of Business Function Steward. We say "steward" rather than "owner" because owners say, "It's mine!" where stewards manage something on behalf of someone else.

Characterize what applications have to do to support each bottleneck removal as user stories and you're about done. Because you're fixing one bottleneck at a time, priority setting is automatic; this greatly simplifies the backlog prioritization that can bedevil a more traditional Scrum team.

Agile COTS and SaaS

The most popular and best-known Agile variants, Scrum and Kanban, describe what software is supposed to do using three levels of focus: epics, which describe something the system should do in very broad terms (for example, *create and edit documents*); features, which list capabilities needed for an epic to be satisfied (for example, *enter text, format text, store documents as files*); and user stories, which describe something specific a user wants to do (standard form: *As a document creator I want to be able to define styles that describe all formatting for selected text, so I can easily format document parts consistently*).

Epics are composed of features; features are composed of user stories.

When you license a COTS or SaaS package, your friendly application vendor has already defined all of these for you as part of writing the package.

So when you're implementing a business change and want to take advantage of the package's capabilities, there's no point in defining your own epics, features, or user stories. They're already embedded in the package.

Which doesn't mean you can just install it, train the users, and be happy. Far from it. You also have to convert and load data and integrate the package with your existing applications and data repositories—meat-and-potatoes activities mostly outside this book's scope.

What we need, and what we're about to provide a sketch of, is an Agile approach to package implementations that don't stop with implementing the package. It's using Agile principles to make the affected parts of the business run differently and better, synchronizing and incorporating COTS or SaaS package reconfiguration into the heart of the implementation practices.

The Agile way of accomplishing this is called CRP (conference room pilot), which has enough similarities to another Agile variant called ATDD (Acceptance-Test-Driven Development) that we consider them interchangeable.

Here's the step-by-step:

Step 1: Hire or train an IT system guru or two. You'll need someone who can make the package you're implementing sing, dance, and play the tuba for what follows to work. While you're at it, work with managers from the affected business areas to assign to the team several employees who are the most knowledgeable in their areas regarding the business functions the new system will be supporting.

Step 2: Installation. Install the package, convert and load the master data, and hook it up to whatever integration technology you use so its data stay synchronized with everything else lying around. IT handles this task on its own.

Step 3: Collect test cases. Either print out a stack of randomly selected business actions processed the old way over the past few months or use your knowledge of such things to create formal test cases.

Step 4: Lock the team in a conference room, along with the test cases and system access. Arrange to have coffee, other caffeinated beverages, and the occasional pizza delivered. We suggest you also establish plumbing allowances—there's no reason to be mean.

If you can't bring everyone together to a single physical facility . . . try harder to make that happen. This all works much better when everyone is face-to-face. But if you can't, that's where ATDD comes in.

Google it for protocols that fit your situation. We aren't going to cover it in depth here, beyond recognizing that it is sometimes an issue. Deal with it as you deal with any other situation in which people who can't be colocated still have to function as a team.

Step 5: Start processing the test cases. Whenever a business user can't process a test case, she explains the problem to one of the system gurus, who then makes use of the package's built-in configuration tools* to make whatever adjustments are needed to process the test case.

Step 6: Loop to step 5 until the application can process all of the test cases, except for those where it makes more sense to kick them out for manual exception handling. For those, document how to kick it out and how to reenter it into the flow of work once it's been handled.

If all goes according to plan, the step 5/step 6 loop will take place in two very natural, overlapping phases. During the first phase, users won't be able to process many test cases at all. Phase 1 is about making the application *competent*.

Phase 2 is about *optimization*—making the business function more effective through the use of the new

---

* A hard-and-fast rule applies here: when processing requires the software to do something it doesn't do out of the box, IT will make use of user-definable data fields, user-configurable business logic, and the package's built-in workflow definition tools, and in extreme cases will program satellite applications that connect to the package through whatever integration technology IT relies on.

IT will never customize the application itself. The long-term cost of maintaining application customizations is hideous.

application. During this phase, users won't be explaining why they can't process test cases anymore. They'll be explaining that if IT could just get the application to do *this* ("this" being a placeholder for a bright idea the business user has), the process would run much better.

This is, by the way, why you need application gurus and the best and brightest business users on the project. Guru-level developers will know what the application is and isn't capable of. Best-and-brightest business users will recognize hidden and correctable inefficiencies.

Synchronizing Agile Methodologies with Strategic Planning

As a reminder, the problem we're trying to solve is that strategic planning generally results in Waterfall-style implementation plans that require information technology best implemented Agile style: iteratively and incrementally.

The most popular solution is SAFe or something like it.

Our thinking: SAFe's proponents are looking at the wrong end of the horse. Instead of establishing all the management overhead you'll need to scale Agile to strategic proportions, rethink how you go about planning and implementing strategy in order to make it Agile (and therefore, agile).

Here's one way of going about it, what we call the 3,1,3,4 approach. By the numbers:

3-year vision: This is what you want your organization to be and to accomplish. You should be able to explain it in clear, direct terms. This isn't the place for nuance.

And if you're a loophole sort of person, sorry, run-on sentences are cheating.

An example, if you're looking for one: "In three years, IT will be the company's partner in designing change and a leader in making it happen."

1-year strategy: This is the one-year down payment on your three-year vision. As is the case for your vision, you should also be able to express your strategy in one or two simple declarative sentences: "This year, our goal is to achieve a 'culture of discipline'—a shared way of thinking and acting that means every employee makes good decisions instead of managers having to enforce them through oversight."

3-month goals: Here's where it starts to become real. Anyone can look out three years, or even one year, and articulate brilliant outcomes. Three months is another matter. Three months is urgent. It's immediate. It's hard to escape.

Developing goals for the next three months isn't particularly challenging. What's difficult is for each department to develop goals that fit together in ways that move you toward your one-year strategy.

Our best advice: Don't worry about it at first. Establishing the habit of putting down any list of three-month goals throughout company management is an interesting enough challenge.

Once the organization is in the habit of setting and accomplishing its separate departmental three-month goals, that's the time to establish a planning process that includes the fit-together requirement.

After that you can start to encourage goals that don't just fit together; they call for active collaboration among multiple departments.

4-week plan: This is where the rubber meets the road. It's what you intend to achieve each week for the next month to make sure your three-month goals turn into accomplishments at a steady pace.

We said "your vision, your strategy, your goals, and your plan," but that's wrong, for two reasons.

The first: if that's all they are—your vision and strategy in particular—you'll fail. Your management team has to embrace the vision; it has to be theirs as much as it is yours.

Your direct reports have to have their own strategies, all of which add up to moving your part of the organization forward toward its vision.

The goals have to be their goals even more than they are your goals, and fan out from strategy just as strategies fan out from vision.

Likewise, "the plan," which is the composite of all the individual plans, should be shared so that everyone knows what anyone knows.

At the end of each month, everyone who owns a plan reviews progress and replans based on how much actually got done and on the extent to which situations and assumptions changed. Likewise, goals should be reviewed on a quarterly basis, and strategies annually.

It's the frequent replanning that makes both the 3,1,3,4 approach and the organization agile: It expects and adapts to change naturally. It broadens ownership, collaboration, and consensus.

And it takes advantage of all of the knowledge about What's Really Going on Around Here that exists throughout the organization.

What's left is checking off the details as the organization completes them. That and all the hard work of making it all happen.

Don't make the mistake of minimizing the hard work just because other people have to do it. Compared with their work, all of your planning is pretty easy.

If You Remember Nothing Else . . .

- Agile is a way of thinking before it's a set of specific techniques. It starts with high levels of direct, developer-to-user interaction and depends on iteration and incrementalism for success.
- The bigger the project, the higher the risk of failure. Most multiyear projects fail, in part due to their complexity and in part due to their design assumptions being superseded by the business changes that have taken place since the project was launched. Success comes from singles, not home runs.
- IT mostly buys, integrates, and configures COTS/SaaS packages, not in-house-developed applications. The best Agile variant for this work is CRP, not Scrum.
- Agile as currently defined and practiced is all about software product delivery, not about achieving intentional business change. Enhancing it to "extend the goalposts" isn't all that hard, but it does take more than just saying the words.

What You Can Do Right Now

- Introduce all Agile project managers, coaches, mentors, and thought leaders to the idea that their jobs are no longer done when the software implementation has delivered all epics, features, and user stories in the backlog. Their projects must include everything required to achieve intentional business change.

- While you're at it, introduce the same group to CRP and/or ATDD as the starting point for implementing commercial software.
- Introduce the company's strategic planners to the concept of Agile business strategy formulation, including the 3,1,3,4 planning framework (3-year vision, 1-year strategy, 3-month goals, 4-week plans) as an alternative to Waterfall-style strategic planning and its intrinsic need to "scale Agile."

4

BusOps

The first rule of any technology used in a business is that automation applied to an efficient operation will magnify the efficiency. The second is that automation applied to an inefficient operation will magnify the inefficiency.
—Bill Gates

No partnership between two independent companies, no matter how well run, can match the speed, effectiveness, responsiveness and efficiency of a solely owned company.
—Edward Whitacre Jr., former CEO,
GM and AT&T

Interlude:
Operations Might Not Be Sexy, but Ignore It at Your Peril

Our medium-sized business recently replaced our core software system, which was built on the client-server model, with a new web-based package. We went about this the usual way: we compiled a list of our requirements, emphasizing business rules that were unique to our situation, then searched for packages that could handle them.

We knew we'd have growing pains trying to configure an off-the-shelf solution to our particular needs. At the same time, we wanted to move to the cloud, as we recognized this is the future and would eliminate the need to purchase, maintain, and manage expensive hardware. And so we created a formal RFP that we sent out to potential vendors.

After reviewing their responses, we homed in on two of the proposed solutions: the ones that looked like they'd adapt to our unique business requirements best and that also were cloud based.

Following our vendor meetings, we called references and talked further about our concerns about configuring the software to our needs. Fully aware of the challenges, we finally chose the vendor that had the best reputation, with the system that appeared to best meet our needs.

As we implemented the solution, we struggled to implement our business rules (no surprise there) to make the system support our business processes. With some pain and lessons learned we reached the punch list stage and checked off every one. Finally, we were ready to roll out our new solution.

That's when we discovered a major problem, one we'd never considered might be an issue. The new system didn't perform as well as our old client-server solution. It wasn't even close. The new system was sluggish. Data entry slowed and our ability to get work done deteriorated.

We were moving to the cloud, so it hadn't occurred to us that we might need an IT operations professional on the team until after we experienced the consequences of not having one.

If you've been involved in a cloud migration, you probably know what's coming. What used to run at wire speeds on our local area network was now running across the internet. Which was when we learned that with the cloud we couldn't just ignore the network when it came to performance engineering.

And so our project team learned about such niceties as net-work latency,* bandwidth,† and what can affect the performance of JavaScript.‡

The project team has since learned these new words (and concepts) and was left wondering how they missed it all.

Meanwhile, I'm proud to report that while in private the head of IT Operations was shaking his head, in public he never once said, "I told you so" or otherwise faulted the project manager for failing to include his team in the project.

He didn't have to. From that point forward his problem was that he didn't have enough staff members to support all the projects that were asking for them.

If There's No Such Thing as an IT Project, Is There Such a Thing as IT Operations?

In the world of organizational design, work comes in two forms: projects and operations.

Projects have a start and a finish. They produce tangible and unique work products that in some way, shape, or form make tomorrow different from yesterday.

Operations also produces work products. Unlike projects, though, the work products operations delivers are highly sim-ilar. In operations, tomorrow is supposed to look a whole lot like yesterday, because consistency is the key to success.

---

* This is the time it takes traffic to move across the internet. Traffic rarely goes from point A to point B. It usually hops around a dozen places before reaching its destination.

† How wide the freeway is in lanes. Add new on-ramps and you may have a problem.

‡ A programming language used to make internet applications feel more interactive.

The premise of this book is that to make the achievement of intentional business-change routine and expected, it must be the planned outcome of what used to be thought of as "IT projects."

Which leads to a logical question: If IT application delivery is just one dimension of a business-change project, does that mean IT operations—the part of IT that makes sure applications, once installed, are running properly and available for those authorized to use them—is just one dimension of business operations?

That's a simple question. The answer, in contrast, is complicated.

IT Operations as a Service Provider

As you're probably tired of reading by now, in many organizations IT is run as if it were a separate business—a service provider for its internal customers. We've already talked about the dysfunction this creates on the applications side of the IT house. It carries over to IT Operations too.

For example, in yet another case of metaphor-driven cures to largely imaginary maladies, the IT-as-a-business metaphor has led to a seriously strange practice: negotiated service level agreements (SLAs) between IT Operations and its internal customers.

Here's the "logic" behind the practice.

When real IT outsourcing businesses negotiate contracts with their real, paying customers, they need a way for both parties to agree that the outsourcer is delivering the services it promised.

And so their contracts include SLAs for each provided service. In general, an SLA is a two-part metric. The first part is the minimum acceptable standard of service. The second part is how often the outsourcer has to achieve that level of service.

In some cases, defining the first part of the metric is trivial, as it is for critical systems where the minimum acceptable standard of service is "up and available." The second part is more interesting; the requirement might be that the system has to be up and available 99.9 percent of the time.

In most situations, though, both parts of the metric must be defined, as is the case for the response to system outages. In this case the SLA might define the maximum acceptable standard for outages when they do occur to be one hour or less before restoration of service. The contractual SLA might specify that the outsourcer must achieve this level of service for 99 percent of all system outages.

For outsourcers and their customers, SLAs are contractual matters. If the outsourcer fails to meet its SLAs, the contract specifies remedies, which are also a matter of contract negotiation. If the outsourcer refuses to provide the specified remedy, the customer can pursue the usual set of escalating legal alternatives.

If internal IT is supposed to act like an independent business, what could be more logical than it negotiating SLAs with its internal customers?

As it turns out, the answer is, lots of alternatives are more logical. Finding some that are less logical? That's the challenge (but one we won't bother exploring here).

Post-SLA IT Operations

Internal SLAs were never a particularly good idea for a number of reasons, some long-standing, some evolving.

The first, already stated, is that they reinforce the wrong IT/business relationship model—that of IT-as-a-business selling to internal customers.

The second is an obvious consequence of the difference: If internal IT fails to achieve a negotiated SLA, what will its "customers" do—sue? SLAs without nonperformance penalties are futile. SLAs with nonperformance penalties encourage interorganizational distrust.

The third is an example of Lewis's first law of metrics: *you get what you measure—that's the risk you take.* In this case IT Operations does measure service levels but lacks any metric regarding innovation.

Well-run IT Operations has to constantly balance between reliability and innovation. But any innovation entails some level of risk. Because SLAs look backward, not forward, they report only the negative consequences of innovation, not its benefits.

As an example, the initial conversion to solid-state hard drives was definitely risky. Their short-term reliability and long-term durability were, for early adopters, unproven. And yet they paid off handsomely for organizations that tried them, giving them a performance advantage in the world of analytics and big data.*

Staying on the leading edge requires some risk taking and forward thinking that SLAs by nature discourage.

Even without these reasons, and even if SLAs once made sense, they don't anymore, for reasons associated with IT Operations' two types of responsibilities: technical services and support services.

SLAs for technical services relate to such matters as system availability and performance. Support services are what the

---

* Example taken from Dave's personal experience.

men and women who work in IT Operations do to provide assistance to the men and women who work in business operations. Support services SLAs relate to such questions as how long someone should expect to wait before the help desk responds to their request for assistance and how long they should expect to wait until the problem they've reported is resolved.

The Case against Technical SLAs

Here's why technical SLAs are (or should be) a thing of the past: once upon a time, high-availability architectures were a choice. Now they aren't.

The fact of the matter is that even when some poor business manager agreed to a technical SLA—98 percent uptime used to be common—he agreed only because he had no real choice. A critical system going down was never acceptable, no matter what the SLA said. Back in the day, the only acceptable service level was the one the telecom department delivered: every time someone picked up the phone they got a dial tone; every time they dialed a number the call went through.*

The Digital world has changed all that. Blame Amazon if you like. When your employees shop there, they never experience an outage, and "employees" includes your executives and managers. One hundred percent uptime all the time, and with snappy performance to go with it, is now what everyone understands is possible, and possible at a scale far beyond anything your company has to achieve.

Oh, and they (and you) also expect that Amazon will process orders, ship merchandise, and even handle returns flawlessly.

_____

* Yes, yes, yes, it's true. With mobile phones we've traded reliability for convenience. But we still get annoyed when voice or data service isn't available.

Service accuracy is baked into Amazon. It's been part of its competitive advantage for years.*

It's now the norm and everyone's expectation.

Should IT Operations continue to track service levels for the technical services it provides? Yes, if it isn't doing very well, but only as a tool to get it to where continued tracking would be a waste of time.

Because while a given piece of equipment might fail, that's no longer a reason for systems to be out of commission. That's the nature of high-availability architectures. If a system is ever unavailable, that should be a sufficiently rare event that keeping statistical track is a waste of time.

What won't be a waste of time is a root cause analysis, because every outage means your high-availability architecture has a design flaw that needs fixing.

What also isn't a waste of time is continuing to analyze reported incidents to detect and address emerging problems early, before they become detectable to the business at large.

Occasional outages used to be a normal part of doing business. In 2018, as we write these words, outages are no longer normal.

Service-Related SLAs

A user has a problem and calls the help desk. The help desk has a service level for time to first response and another service level for time to resolution.

---

* When we use Amazon as an example, we're talking about Amazon the retailer, not Amazon Web Services, its cloud-provider division, which ironically enough does have SLAs in place for its customers. But then, those are real, paying customers, so traditional service models and lawyers are both pretty much inescapable.

On any given day, for any given call, the help desk either re-sponds more quickly or responds less quickly than the service level specifies. It responds more quickly when help desk staff capacity exceeds the call volume. It responds less quickly when the call volume exceeds the help desk staff's capacity.

The SLA has absolutely nothing to do with the help desk's performance. It's just a stick. It's useful for beating up the help desk manager and not much else.

That's almost an entirely accurate statement. The only time it's inaccurate is budget season, when the help desk's service level performance can be used to justify hiring more staff.

This is, to be fair, no small matter. But it justifies the practice of tracking service performance, not negotiating SLAs.

What to do instead of negotiating SLAs: It's as we just said—by all means track performance. Otherwise you won't know what needs attention and what doesn't. It's the negotiated SLAs that belong in the dustbin of outmoded management practices.

The Only IT Operations Metric That Matters

Pity the poor IT Operations manager.

For most people in management, success increases their vis-ibility, which can lead to promotion, accolades, and better pay as well. The only time IT Operations is visible is when some-thing goes wrong.

All good metrics are numerical representations of qualitative goals. And so the IT Operations metric that best reflects its goals is a measure of its invisibility. This "invisibility index" should be a composite metric that encompasses application availabil-ity and performance, the number of calls to the help desk—fewer calls means more invisibility—and (we're finally getting to the point of this chapter!) some measure that reflects how often IT Operations performance is a bottleneck in other areas' business processes and practices.

Fixing IT Operations for IT

Typical IT organizations are divided into Applications and Operations—Apps and Ops for short.

In these typical IT organizations, Apps and Ops distrust each other. One reason is that Apps succeeds by making application changes, but for IT Operations, each application change creates a risk of increased visibility.

A second reason is that Application Development teams need IT Operations to create and maintain development and test environments. For Apps this means Ops is a bottleneck. For Ops this means additional and often unscheduled work.

Which is how DevOps happened. DevOps is a form of Agile (see chapter 3). Unlike most Agile variants, with DevOps, as its name implies, Application Development teams include one or more IT Operations staff to handle IT Operations project responsibilities collaboratively and on the project's schedule, instead of their being handled through the IT Operations request queue.

DevOps has a number of other interesting characteristics, all of which we're going to ignore here as they aren't directly relevant to what this book is about.

Digital Business and the End of IT Operations

It's possible there's never been a management fad more confusing and ambiguous than Digital this and Digital that.

Behind all the ambiguity are specific Digital technologies that create new capabilities. Businesses can take advantage of them to create competitive advantage. Or, they can ignore them, letting competitors gain the advantage instead.

Behind the specifics is the central Digital reality: information technology is no longer optional. It's deeply embedded in every business process and practice your company relies on to do business on a day-to-day basis.

A logical consequence: conceptually, IT Operations is just one collection of moving parts among many in overall Business Operations. It's logically just as much the province of the COO as it is of the CIO.

Digital Business and the Beginning of BusOps

As a practical matter, wherever it reports, IT Operations should remain intact. Its effectiveness (and consequent invisibility) depends on the ongoing collaboration of a number of technically proficient specialists—practitioners of mature and well-developed disciplines.

Managing the processes and practices they're responsible for depends, in turn, on managers who understand their inner workings. Leading them depends on managers who can empathize with these practitioners as they go about handling their responsibilities.

It's also worth recognizing that reorganizations rarely fix anything. Mostly they remove barriers by putting groups that didn't work well together before the reorganization under common management.

Which also means most reorganizations create barriers between groups that used to have common management but don't anymore.

From our perspective, moving IT Operations in the organizational chart so it reports to the COO is neither more nor less logical than leaving it where it is. As for restructuring IT Operations, breaking it up and parceling out its responsibilities within the business . . . that just won't work. There remains value in technical people working together on common problems.

What does need to change? DevOps points the way. The culture of collaboration we discussed in chapter 1 has to extend to the relationship between IT Operations and the rest of business operations just as surely and deeply as it does between business managers who want to run things differently and better and IT Applications.

So unchain your help desk staff. With no SLAs to shackle them to their chairs, you can encourage them to get up and visit someone with an issue, learn about what their challenges are, and offer insights into how else they can take advantage of the technologies at their disposal.

Meanwhile, as you're fixing Agile, fix it more by adding the DevOps dimension of including system and security administrators on business-change project teams. Your projects will have better outcomes, and the additional knowledge of what matters in business areas will make IT Operations more effective in its day-to-day decision-making.

Let's introduce a new term to make it official. Just as DevOps is the blending and collaboration of IT Apps and IT Ops, let's start talking about BusOps* as the blending and collaboration of IT Operations and Business Operations.

The battle, according to military theorists,[1] is always for the hearts and minds. That often starts with vocabulary.

So BusOps it is. Add it to your working vocabulary. You just might be surprised . . . in a good way . . . at what comes of it.

---

* Pronounced 'Biz-Ops.

If You Remember Nothing Else . . .

- Service level agreements (SLAs) are contracts between suppliers and customers. That's no longer the relationship between IT and the rest of the business, so SLAs have to go. Tracking service performance? That's still necessary because otherwise you won't know how you're performing.

- Another reason SLAs are pointless: the days of computer systems going down are over. We live in a 24x7 world. Amazon and Google don't go down, and every executive, manager, and employee in your company knows it. So they won't be satisfied with anything less from IT.

- Business operations and IT Operations have a great deal in common. More than that, they're inextricably linked: the point of IT Operations is to support business operations; without one you can't have the other, and without the other you don't need the one. Welcome to the world of BusOps.

What You Can Do Right Now

- Eliminate all SLAs. Don't worry. The business managers you've negotiated them with will breathe a sigh of relief.

- Recognize, and make sure everyone within IT and throughout the business recognizes, that unplanned down-time is no longer normal or acceptable. Anything short of 100 percent availability means something needs fixing.

- Institute the only IT Operations metric that matters: the invisibility index.

- Introduce everyone in the business, both inside and outside IT, to the idea of BusOps and make it part of everyone's vocabulary.

5

Business-Change Governance

*We'd all like to know how the information revolution
eventually shakes out. I think we know. It ends
in ordinariness, disappointment and advertising,
as things always do.*
—Tom Toles, Washington Post

Interlude: The "Logic" of Political Engineering

*Some years ago, I found myself involved in a would-be project that
stood at the crossroads of a major medical practice plan and its hospital system partner. Both parties had agreed on the need to streamline
the online system that allowed patients to select their doctors, but
neither side demonstrated any willingness to make the structural
changes necessary to make it happen.*

*The problem was that patients had access to two distinctly differ-
ent websites that offered the same doctors at the same clinics. Each site
relied on its own database and middleware, and any consistency between the two systems was essentially a coincidence. Each organization believed the other's solution to be the inferior one. The hospital
system wasn't about to give up its scheduling system, and the practice plan insisted on control over the physician listings.*

Now, I can't think of a more worthy business case than "help patients in need get the right medical care." Unfortunately, the IT brain trusts at both organizations were thoroughly entrenched, and it really didn't matter how simple or laudable the goal was. Absent a champion who both respected the business need and comprehended the means necessary to attain it, this idea was dead on arrival.

My role in the erstwhile initiative was to break the news to the stakeholders that the technical barriers were just too great. In this case, the stakeholders were a group of world-renowned physicians who were unaccustomed to being told what was and wasn't possible.

At the time, I believed I received this assignment because of my ability to explain complex issues to important people without talking down to them. In retrospect, I was probably just sent in because I lacked the savvy to make it somebody else's responsibility.

When the day came, I walked into a room full of highly respected physicians who made no attempt to conceal that this meeting represented an interruption to their important work with patients. I walked carefully through the reasons why their perfectly logical request had no chance of being fulfilled. At the time, I felt as though my explanation was exceptionally clear and well measured.

When I finished, I was met with an excruciating but very brief silence. The words that broke it came from a particularly irate oncologist, and I doubt I'll ever forget them:

"We cure #@$&ing cancer, and you came over here today to tell us you can't get two databases to talk to each other!?"

Culture change encourages everyone to view their world from a different perspective—that just delivering technology won't do the job. The new business/IT conversation makes the culture change real by redefining responsibilities. Fixing Agile gives us specific techniques for making the changes happen, and by instituting the concept of BusOps, changes achieved through Agile techniques have a chance of being incorporated into the day-to-day work of the organization.

Which brings us to the challenge of figuring out how the business decides which business changes to invest in and how much to invest in them. That's what change governance is about.

Start with how the new business/IT conversation starts, with the question, how do you want your part of the business to run differently and better?

The change governance practices you'll need if you're going to succeed in intentional business change . . . in eliminating IT projects in favor of business-change efforts . . . start with the answers to that question.

They end with decision-makers abandoning return on investment—the ever-popular ROI—as the alpha and omega of what constitutes *good* in business.

Not ROI as a concept. That can stay. It's ROI as it's usually conceptualized—as a purely financial metric—that's just too one-dimensional to apply to ideas for improving how business gets done.

To understand why, we have to dive into two related topics. The first is the nature of business benefit, which happens only when one or more of four "goods" improve. The second is the difference between what a business-change project can plausibly deliver and its impact on the four goods.

The "Four Goods"

In even the most enlightened executive suites, only four business improvements really matter: more revenue, lower costs, better-managed risks, and enhanced mission fulfillment.

Three of the four sure do look like purely financial metrics, don't they?

They do. And yet, in practice it doesn't exactly work out that way. Here's why.

The Cost Case

Imagine you propose a project to streamline a manufacturing process you're accountable for. Your proposed improvements will reduce the incremental cost of the doohickeys your company sells by 10 percent. Some spreadsheet work computes an ROI of 20 percent.

It's a slam dunk.

No, it isn't, because you aren't planning to lay off 10 percent of the manufacturing staff. You're going to keep them where they are on the grounds that this way the company can absorb the growth in volume to be expected when the company reduces the price it sells doohickeys for.

Or, you could lay off the employees to make the cost reduction tangible. That lets you lower prices, which increases sales volume, which in turn results in considerable time and expense wasted finding employees to replace the ones you just let go.

Or, you could avoid these complications. You could lay off employees and keep the price of a doohickey the same. Instead of revenue growth you get better margins for the revenue you do get.

But on the other hand, if you go for the margins you don't get improved marketshare, mindshare, and walletshare— revenue's leading indicators.

The investment you're asking for is counted in tangible money. The growth you're basing your plans on is just a promise, and without it you'll have investment without return.

Revenue Reliance

Cost reduction sells. Revenue growth doesn't.

It isn't that nobody wants the revenue. Of course they do. But two major barriers separate the desire for more revenue from the steps needed to get it. The first is that the causal connection between any one action the company might take and increased revenue is uncertain while planning.

The second is that it's impossible to prove following execution.

It's uncertain while planning because unlike cost-reduction projects, which can directly reduce costs (but see above), multiple factors affect revenue, many of which are beyond your project's influence, let alone control.

Among them are how your products compare with what competitors bring to market, the effectiveness of your advertising, the caliber of your sales force, your ability to build and deliver products compared with that of your competitors, and how well customer service in all its forms minimizes customer defections.

Revenue is multivariate. When it changes, whether it improves or deteriorates, proving what caused the change is close to impossible. That being the case, in ROI-driven businesses, revenue-based business cases are generally viewed with skepticism.

Which is why, in ROI-driven businesses, given a choice between a revenue-enhancement project and a cost-reduction project, cost-reduction usually wins.

The Risk of Risk

A risk is something bad that might or might not happen. This is different from the risks actuaries price. Actuaries deal with statistical risk. They're dealing with the certainty that given enough cases, the something bad in question will happen a certain fraction of the time.

Actuarial risk results in predictable costs, which is why a program to improve employee health would count as cost-reduction, not better risk management: while on any given day a specific employee either is or is not ill, over the course of a year and across all employees, the total number of sick days taken is quite predictable.

Contrast that to, say, a proposed investment in next-generation information security—an investment in anticipating new types of threats and deploying preventive responses. Further imagine that for many of these next-generation threats, you'll have no way to detect unsuccessful attacks.

And so the company decides to approve the proposal. Five years later your systems have never, to the best of your knowledge, been penetrated.

Does this mean your project was a success, or does it mean your systems were never attacked?

The answer: successful prevention is often indistinguishable from the absence of risk, and the less an audience wants to understand the details, the more this is true.

That's the problem with investments in risk management. There's rarely a way to attach a reliable probability to a given risk, and without that probability there's no way to compare the value of prevention, mitigation, or insurance with their costs.

And so, with ROI as your company's metric, risk-management projects have a hard time competing for funding with those that promise cost reductions.

A complete discussion of this topic is beyond the scope of this book. Here, it's worth mentioning that the world of business has made significant progress, specifically because when it comes to enterprise risk management, boards of directors and members of the executive suite no longer insist on ROI to justify investments.

Instead, the conversation is about whether a given risk is plausible enough to warrant investment in a response, and if so, which of the four types of risk responses makes the most sense, the four responses being:

1. **Prevention** (also known as **Avoidance**), reducing the odds of the risk becoming reality
2. **Mitigation**, reducing the damage if the risk does happen
3. **Insure**, spending a fraction of what the risk would cost if it became real now, so as to share it out if it does
4. **Accept** (a version of which is **hope**), which is how businesses deal with the possibility of an asteroid striking the earth because Bruce Willis didn't get to it in time

The Merit of Mission

With or without a direct financial business case, your organization's mission is a perfectly valid investment target.

Not what your mission statement talks about. Forget mission statements. They're worthless. Understanding your mission, on the other hand, is vital, once you understand that "mission" means what it is you do that provides value to your customers, and what it is you do that provides enough unique value that your customers decided to be your customers instead of spending their money elsewhere.

What mission doesn't always do is directly drive profits. Take, for example, the pre–Great Recession General Motors. Its neglected mission was to sell cars people wanted to buy. Most of its profits came from financing the cars it sold, which explained how it came to be that GM eventually had to bribe consumers to buy its uncompetitive cars through the expedient of providing sizable rebates.

And even with the bribes its marketshare eroded.

As a second example, look at the business model followed by print media and most cable channels. To the unenlightened,* when they watched *Sharknado* on the SciFi channel, they were watching one of the SciFi channel's products.

But that wasn't what was happening. *Sharknado*, the daily newspaper, and other such things aren't products. They're bait, designed to attract media consumers who the media company in question then sells to advertisers.

From the perspective of most media companies, we're the product.

In the long run, businesses can't succeed when they fail at their missions,† but the delay between mission erosion and business failure can, as was the case with General Motors, be long enough to create the illusion that mission success and business success are uncorrelated.

The business model most of us learned about first was simple and straightforward. Call it the "better mousetrap model." It's the model followed by companies that make something tangible (the mousetrap), selling it for more than the cost of making it. Investments in mission are easy to justify when your business adheres to the "better mousetrap" model.

And it's why investments in mission are nearly impossible to justify otherwise: the costs are immediate, but the benefits are delayed, indirect, and possible to demonstrate only after years of making the wrong decisions.

---

* Those who haven't read this book.

† This includes those companies that mistake their eminently ignorable mission statements for their mission.

Can't We Fix Problems?

Sometimes, the rationale for a project is that it fixes a problem. Isn't that good enough?

Well . . . no. If there's something or other that's annoying but doesn't have any impact on revenue, cost, risk, or mission, it's a darned* shame. But the world is filled with annoyances. That doesn't mean it makes sense to try to fix them all.

Annoyances rise to the status of "problem" only if they have some impact on one or more of the four goods. They might be too small for the impact to be measurable, which is okay if the cost of the fixes is also too small to be more than a rounding error. Otherwise, the problems in question probably aren't worth taking the time and effort to resolve.

But Wait—What about Strategy?

Oddly enough, business strategy doesn't enter into change governance very much. It isn't that strategy doesn't lead to change. Of course it does. It's that business-change *governance* focuses on evaluating and supporting bottom-up proposals for business improvement.

Strategy—its definition, its funding, and its implementation plans—comes from the executive suite. The business-change governance practice does not have a role to play in evaluating it.

Value Levers

Value lever analysis isn't, we're sad to report, original with us. The concept is, however, essential to understanding how business-change governance should work.

-----
* Sorry for the strong language.

While there's no one model that fits all businesses . . . and no one model that's standardized across all consultancies either . . . they all work something like this:

- **Tangible change:** what's going to be better tomorrow than it was yesterday, with "better" usually defined by the six dimensions of optimization as described in chapter 2.

  While not universally true, most tangible changes will be quantitative and measurable—for example, the change is expected to result in a twenty-minute (10 percent) cycle time reduction.
- **Business outcome:** the direct reason you care about the tangible change you're trying to implement—for example, improved customer retention, increased customer walletshare, product superiority, or improved product pricing.

  Because most business outcomes are multivariate, affected by more than one business change, the estimated impact on each business outcome will usually be qualitative.
- **Four-goods impact:** how the business outcome will affect revenue, cost, risk, and mission.

  The connections between improvements to revenue, cost, risk, and mission and the business goals that drive them are multivariate. The connections among improvements to business goals and the projects that drive them are also multivariate. Consequently, project proposals should establish their logical connection to improvements in one or more of the four goods without making promises of provability.

You might be tempted to try to connect tangible changes to business outcomes and business outcomes to the four goods quantitatively. With sufficiently excellent analysts and carefully chosen analytics, this just might be possible.

But we're skeptical, and even if a sufficiently brilliant analyst managed the modeling required, you'd need a way to test the model before you could rely on it. You'd also need procedures for updating the model after every business change.

Qualitative linkage, accompanied by clear explanations of why the linkage is logical and plausible, usually has to do, and is usually sufficient for, governance purposes.

With the above in mind, here's a framework for business-change governance to get you started.

Business-Change Governance

Effective business-change governance depends on the culture change we described in chapter 1, specifically:

Where there are IT projects, business executives view project proposals with suspicion. They see their job as screening out bad ideas, which is why they insist that the CIO provide a hard-dollar return on investment, typically calculated as the number of warm bodies to be laid off multiplied by their annual compensation plus benefits.

Where there are no IT projects, business executives view project proposals as opportunities to improve how their company conducts its business. They see their job as helping good ideas succeed. Because of this they insist that all project proposals be described in terms of business change, described that way by a business sponsor who is enthusiastic about the possibilities and supported by an IT SME who attests to the project's technical feasibility.

Effective change governance is about helping good ideas succeed.

Which in turn means that whatever it's called and whoever participates, before the change governance body* evaluates its first project proposal, it:

1. **Obtains the business-change budget.** Determining the funding available for business-change projects is the first business-change governance decision. The rest of business-change governance is deciding how to allocate it. Speaking of which, in steps 2 and 3 the council allocates the budget.

   When preparing the budget, all parties need to take the distractions of the present into account when reserving staff time for making the future get here. This is especially true for the men and women you want to include in business-change efforts the most, because they're the ones who are likely to be called away to handle day-to-day urgencies.

2. **Reserves a slice of the pie for enhancements.** Enhancements are small items—something a programmer can handle in a couple of weeks but certainly no more than a month. A manager might need a new report; moving a field from one screen to another might make life easier.

   The council should exclude itself from decisions about specific enhancements, as this is the business equivalent of being pecked to death by ducks. Instead, it allocates a pool of IT developer hours to each business area for its use in requesting enhancements.

---

* Call it the Business Change Governance Council, as explained in the next section.

As an alternative, maintain an Agile-style backlog* for each application. The size of each application support team, set by the council, determines overall capacity. Managing the backlog is a process/product owner responsibility.

There's a principle of information theory that applies here: one person's signal is another one's noise. From the perspective of the CIO, taken collectively, the enhancements queue is noise. From the perspective of the requesters, it's a very important collection of signals.

3. **Allocates** the total business-change budget, either in dollars or in project-team-member hours, among the four goods, so that, for example, individual risk-reduction projects don't compete for funding with individual revenue enhancement projects.

4. **Establishes and ranks** the company's most important **business goals:** customer retention, product innovation, and so on.

5. **Publishes guidelines** for the ingredients of a successful **proposal**, which are:
   - *Change stewardship.* If the point of a project is to improve a business function, the change steward, who corresponds to the business sponsor in traditional environments, and product owners where Agile holds sway, is the manager or executive directly accountable for the affected business function.

     The inviolable rule: no change steward, no project.

---

* For those who aren't Agilistas, the backlog is the list of desired capabilities for a new software application under development. It's up to the product owner to decide their relative priority. We figure, just because an application already exists, this doesn't change very much: in an Agile project, after a month or so the core application exists too.

- Intended business change:
  - Which business function or functions will improve.
  - For each affected function which of *the six dimensions of optimization* will improve and, preferably, how much they'll improve.
- Which *business goals* will be closer to achievement and how the planned business function improvement will bring them closer.
- Which of the *four goods* will improve, based on the value levers analysis the company's strategic planners have already performed and published, or, if they haven't, how your own four-goods analysis connects the relevant dots.
- Connection to the company's *strategy and strategic plan.* Note that there are plenty of good ideas that deserve to be pursued that have no direct connection to the strategic plan.

  That's okay, and in fact it's commonplace for localized, nonstrategic projects to increase revenue or decrease costs in ways that can help pay for the strategic plan.

  What aren't okay are projects that run counter to the strategy, that make accomplishing the strategy more difficult, or that risk dragging the company in a different direction.
- *Compliance impact.* Documentation that the change design conforms to whatever compliance regimes are relevant to the change in question.
- *The plan.* Preparation of a detailed project plan in advance of receiving approval to proceed would be a waste of time. Failing to have any plan at all, on the other hand, suggests a lack of seriousness—an expectation

that daily improvisation coupled with good intentions will somehow lead to successful implementation.

Somewhere between these two extremes is a road map that describes the most important work streams, lists the project roles needed for each of them to get the work done, and puts them all on a timeline.

- *Ripple effects.* Most changes will, in addition to their desired consequences, affect other parts of the organization.

  At a minimum, business changes that require significant investments in information technology will probably need ongoing support. The organization needs to be prepared for and agree to fund this cost.

  Beyond this, it's likely that changing how one business function goes about things will have some impact on other business functions it's connected to. These are all part of the business change the proposers are recommending. If those responsible for these ripple-effect-affected business functions haven't agreed to participate, the proposal isn't complete.

6. **Evaluates proposals.** With this foundation the governance body is ready to evaluate proposals for business change. To do so it:

   - **Consults** with proposers whose proposals are weak but seem to have good ideas hidden inside them, to help them improve their proposals. The goal, you'll recall, is to help good ideas succeed, not to choose the best-written proposals.

     This is very much parallel to the hiring process, where smart managers recognize the difference between the candidate with the best résumé and the best candidate.

- **Decides** which proposals to approve. Along with this, a tip: once the council reviews a proposal, decision-making is binary. The two possible results are *scheduled* and *rejected*.

  There is no *yes* that means anything unless the council has added the project to its master schedule, which means the project has a start date, on which the staff needed to execute the project will be available and ready to begin.

  Also, there's no *maybe* or *not yet*. These are fictions. If a proposal isn't good enough to be put on the master schedule now, it never will be. Don't pretend.

  In companies that violate this rule, their change governance body is inevitably and perpetually plagued by approved proposals that never seem to launch. Which is why the two outcomes for any proposed project are *scheduled* and *rejected*.

- **Ensures** all approved projects are fully staffed. Fully staffed means the project never has to wait for the employee responsible for a project task to become available to work on it. We're stealing a page here from Eliyahu Goldratt's Critical Chain methodology, which teaches that it's far better for an employee to have idle time than for a project to sit idle because the employee isn't available.

  And no, multitasking doesn't solve this. All it does is provide a euphemism for interrupting interruptions with interruptions. If two tasks take twenty hours each when performed sequentially without interruptions, you can bet they'll take at least fifty hours in total when an employee has to multitask to handle them concurrently.

Full staffing is the most important reason to make a master project schedule central to the actual work of business-change governance, as it is the tangible mechanism through which staff commitments can be managed.

7. **Reviews progress**, for projects that are in flight, and outcomes, for projects that completed since the last review period.

Projects get into trouble for any number of reasons, a small fraction of which are bad management and problematic project teams. Vendors don't deliver, employees leave the company, sponsors receive promotions . . . the list goes on. Troubled projects need either help or cancellation. The council is responsible for both.

As for outcomes, a project proposal describes a project's promise. The council evaluates the proposal to decide how likely it is that the promise will pan out. This means the council needs to determine whether it actually does, not to "hold the proposer accountable" but to evaluate and improve its own competence at evaluating proposals.

The Business Change Governance Council

Companies that have IT projects typically have an IT Steering Committee to govern them. By now it shouldn't be a stretch that as we're moving the goalposts from IT product delivery to intentional business change, it's no longer IT that we have to steer.

What's less obvious is something else we don't need: a committee.

But it should be obvious. Few would argue with the proposition that businesses need to be faster and more nimble. And yet, equally few recognize one of the immutable truths of organizational dynamics: no matter how slowly things are happening now, you can count on a committee to slow things down even more.

Which is why we recommend forming a council instead. As we use the terms, a committee's members *represent* the part of the company they come from. A council's members, in contrast, see themselves as *leaders of the whole enterprise*, who just happen to currently focus on the area they're responsible for.

A committee's members negotiate and trade favors. A council's members collaborate to figure out what's best for the company as a whole.

Two cautions: first, even with the best of intentions, even the most council-oriented governance group can deteriorate into a committee. Changes in membership are a particularly sensitive situation, as new members are likely to walk in the door with the expectation they're joining a committee.

Second: all it takes is one "cheater on the system"—a council member who starts to advocate for their silo—for the council to start the short, quick slide into committee-ness.

Shadow IT

Back in the early days of personal computers, much of their appeal was that users didn't have to wait for IT to get around to their request (IT never did because it was already eaten alive by bigger ones); instead, users could DIY a solution quickly, affordably, and without IT even knowing it had happened.

They might push spreadsheet software to its limits. They might build a database using one of the several end-user DBMS packages designed for this, like dBase II, Paradox, or Access.

They might buy and install an inexpensive commercial package. Sales reps, for example, loved Act! because it was (1) cheap,

(2) easy to use, and (3) designed to help sales reps sell. It was the ancient-day equivalent of installing a smartphone app.

IT had, and mostly still has, a hate/hate/hate/hate relationship with shadow IT. IT hates it because from time to time, shadow IT-ers need help, and IT isn't in a position to provide it. It hasn't, after all, been involved in implementing the shadow system, so it has no expertise to offer.

IT also hates shadow IT because it makes IT look bad: end users make their own IT happen, successfully, and at a fraction of the time and expense IT would have needed to accomplish the same result.

IT has legitimate concerns as well, especially with respect to security and compliance, as well as support and integration. Not that IT has a perfect track record when it comes to these critical challenges, but at least it has processes to implement proper controls for security and compliance, governance practices to plan for support, and, to some extent, architectural standards for integration.

And finally, IT hates shadow IT because it can't stamp it out. After all, stamping out shadow IT means IT saying to its business customers, "We won't do it for you and we won't let you do it for yourself."

Especially for those IT shops that adopted the internal customer metaphor,* telling business users they can't implement their own IT is tantamount to Home Depot refusing to sell drywall and joint compound to shoppers.

And yet, one way or another, most IT advisers recommend locking down desktops and keeping end-user DBMSs out of the hands of end users, the business case being recognition of all the disadvantages while carefully ignoring all the benefits.

---

* All of them.

Which lasted until the advent of the cloud and Salesforce. IT could lock down desktops. Locking down the internet proved to be quite a bit more difficult.

The Solution: "Illuminated IT"

As there's no longer any such thing as an IT project, it stands to reason there's no longer any such thing as shadow IT.

Quite the opposite: business DIY projects have the advantage of business change and improvement being the whole point. IT can come by later to replace the inadequate plumbing and wiring so they meet its metaphorical building codes, without having to be very involved in designing and implementing the business change the technology is supposed to enable.

Something else IT can do is to provide the integration that's almost always impossible for shadow IT projects. Integration is arguably IT's most challenging responsibility, as the analysis and technology needed to interface systems with different origins and design philosophies is intrinsically complex.

It's like this: a sales manager can sign up for Salesforce licenses for every sales rep in the company. When data entered into Salesforce are needed in a different system, though, someone has to rekey them from a Salesforce screen into the other system.

Unless and until, that is, IT steps in to provide the integration.

IT has other expertise business users generally lack—not only integration but also methods for choosing the best package or tool and managing the project itself. Beyond this, if a business DIY change project neglects to pay attention to relevant compliance requirements, IT can provide a backstop.

And an oddity: often, shadow IT teams forget that the point is to change how business gets done. They become nothing more than additional IT teams only without IT's special expertise. In the new world order we're advocating in this book, IT can provide consulting that helps illuminated IT teams keep their focus.

How to do this: usually there are employees throughout the business who have skills and aptitudes comparable to IT's business analysts (now internal business consultants). Give them all the education, advice, and other support they can handle, and they'll do much of the heavy lifting required to make illuminated IT projects . . . strike that, illuminated *business-change projects* . . . successful.

Illuminated IT can greatly expand a company's capacity for business change. With IT's support it's possible to minimize the disadvantages while reaping most of the benefits.

Illuminated IT requires a fundamental change in how IT works and thinks. It is, in fact, the exact same fundamental change this book is about: IT no longer "owns" the technology silo. It's a technology steward, but it's a partner and collaborator with the rest of the business, in this case supporting business-led technology implementations as part of its role in enabling intentional business change.

In Conclusion

Change governance has quite a few gears and pulleys, and there's no one way to connect them that fits all situations. We've tried to provide useful guidelines. Your success will depend on experimentation—a willingness to try new techniques and, even more important, to admit when something you've tried isn't working out as hoped.

Behind it all, and most important for change governance to be the linchpin of intentional business change instead of its greatest impediment, is the difference between committee and council.

If your change governance body is composed of people who see themselves as leaders of the whole organization and not representatives of one of its parts, the rest will happen.

If not, you'll probably see some forward progress, but it will be nothing like the company's actual potential.

If You Remember Nothing Else . . .

- Investments in business change should benefit one of the "four goods": increased revenue, decreased costs, better risk management, or improvements in accomplishing the organization's mission.
- The group responsible for business-change governance should be constituted as a council, not as a committee, the difference being that council members consider themselves leaders of the whole organization, currently responsible for a chunk of it, as opposed to committee members, who consider themselves to be representatives of the silo they currently lead.
- Business-change governance should be devoted to helping good ideas succeed, far more than screening out ideas that aren't worthwhile.
- Business-change governance reviews should have only two possible outcomes. Proposals should be either scheduled or rejected. If a project isn't important enough to be placed on the project master schedule, it hasn't really been approved, and pretending has no place in effective business-change governance.
- Shadow IT—DIY business-change efforts, with DIY including the implementation of information technology—used to be preventable. With the advent of SaaS, IT can't stop it

anymore, nor should it. Encouraging and supporting it turns shadow IT into "illuminated IT," greatly increasing the organization's change bandwidth.

What You Can Do Right Now

- Disband your IT Steering Committee and replace it with a Business Change Governance Council. Make sure everyone involved embraces the difference between a committee and a council, and commits to it.
- Build a project value framework around the "four goods" (revenue, cost, risk, and mission), making use of value lever analysis as the means for connecting project results to them.
- Establish an "illuminated IT" support function within the IT organization and publicize its availability and capabilities throughout the rest of the business.

6

IT in the Lead

You do not lead by hitting people over the head.
—Dwight Eisenhower

Interlude: Going Paperless for Real

We run a complex business. More to the point, our business has a complex sales process that relies on a global network of three hundred independent manufacturers' representatives who custom design solutions for their customers' manufacturing lines.

It's a highly competitive business that creates a steady stream of new products and specification changes to meet customer demand. The manufacturers' reps need detailed, readily accessible information about each component we sell to help them design systems for their customers.

And did I mention they're independent? They're free to source components from us, our competitors, and often both.

To support the manufacturers' reps, our Sales Support organization printed and mailed out product catalogs in three-ring binders. Each new product introduction produced several new pages of product detail. Sales Support mailed or emailed out new or replacement pages.

The reps had to constantly update their product catalogs, pulling out old pages and inserting new ones.

I run IT. When I received a request from Sales Support to select and implement a new printing and delivery system to more efficiently provide catalog updates to manufacturers' reps, I wanted to satisfy my internal customer, so I put my best business analyst on the hunt.

Meanwhile, we were already in the middle of satisfying a different request, in which we were rolling out iPads to our traveling employees. The pilot group found the iPads to be handy for email and web browsing. However, adoption was tepid, since iPhones were already doing the same job.

Fortunately for all of us, IT's business analysts and project managers talk with each other, which is how it happened that the business analyst I'd put in charge of the new catalog printing and delivery system and the project manager responsible for the iPad rollout dropped into my office.

"Kill the catalog printing and delivery system project," they told me. (I'd say "recommended," but really, they told me.)

Instead of an IT project based on requirements drawn up by Sales Support, they assembled a team from Sales Support, the manufacturers' reps, Marketing, Engineering, and IT. They proposed to solve the printing and distribution problem by not printing and distributing anything anymore. Instead, our so-called internal customers became our design collaborators in the Slimline Product Catalog initiative. The easy-to-sell concept was that manufacturers' reps could avoid hauling around heavy product catalogs, replacing them with an easier-to-carry electronic version designed for the iPad's form factor and capabilities.

Along the way the team adopted an Agile approach, not that they called it that or cared. Their new application development process repeated a cycle of brainstorm, prototype, test, critique, and modify, fine-tuning the app and adding new functionality with each iteration.

Our planned benefit was a dramatic reduction in printing and delivery costs, a fact we didn't once mention to the manufacturers' reps. Our message to them: the new tool was an easy way to look up parts and their specifications more quickly than paging through a five-hundred-page three-ring binder, and oh, by the way, they wouldn't need to waste their time replacing pages over and over again.

When the app was first rolled out to a pilot group of manufacturers' reps, it contained only basic search and display functionality for all SKUs in one simple product line.

Our reps loved it.

After the fifteenth development iteration, Slimline was providing three hundred manufacturers' reps with an up-to-the-minute product catalog with features far beyond those provided by the print catalogs: capabilities like search/filter by product family, application, part number, specifications, competitor's part cross-reference, availability, and price. It delivered multilingual capabilities, product drawings, order entry, shipment tracking, and returns processing.

As we added incremental features, we discovered our original business case had become just a fringe benefit: while the original print-and-distribute request was justified by projected cost-cutting, it turned out our electronic catalog resulted in increased revenue, as the manufacturers' reps were now specifying our products over the competition more often, because it was a lot easier to do so.

Oh, and we did substantially reduce our printing and distribution costs too.

One more thing, where before this project I had a hard time making IT anything more than an order taker, after it, we started to get requests that weren't just for information technology.

They were to help people figure out better ways of getting things done.

The previous five chapters have set the table. They've established what organizations need to do to achieve excellence at intentional business change, including the means for sorting through the large stack of business-change proposals that is a feature of healthy organizations.

But not all proposed changes come from within the organization. Those in the executive suite are, after all, paid to lead the business, which includes responsibility for setting and planning strategy.

When there's no such thing as an IT project, this changes too, and in ways you might find startling, starting with the CIO's role in the business.

By now the idea that the CIO should "have a seat at the executive table" is well-explored territory. What's less clear is what the CIO should do once seated there. A common but flawed version is that the CIO should sit there to hear what everyone wants firsthand instead of at a distance, and to hear everyone's gripes about IT's performance so that IT finally takes the gripes seriously.

Here's what it should mean: IT leads planning. This one phrase has three very different but equally important meanings. The first is leadership in planning—helping everyone who needs the help understand how to turn good intentions into a plan of action. The second is driving much or all of the company's strategic planning. The third is serving as the gateway to business-change governance.

Leadership in Planning

Role Reversal

IT and manufacturing are in the process of reversing roles.

A lot of the past conventional wisdom about making IT work right was based on making it more like manufacturing. Writing software was supposed to become a factory-like situation . . . never mind that the goal of manufacturing is to make lots and lots of products that are as identical as possible, while the goal of IT is to write lots and lots of software modules, each of which does something entirely different from all the rest.

But never mind all that.

In typical businesses, most executives' and managers' expertise is in keeping the joint running—in making sure the business runs the way it's supposed to run, to produce, sell, and account for the products and services its customers buy and pay for.

Their expertise, that is, is in maintaining and perfecting the status quo.

But we're living in a world of change that isn't just constant; it's accelerating. And while nobody in the executive suite is fully versed in making change happen,* the CIO, who represents all of IT's capabilities by proxy, is, in most companies, better versed in making change happen than anyone else.

IT is in the change business. By providing leadership in change planning, it will make manufacturing, and by extension the rest of the business, more like IT.

Including the CIO on the executive leadership team (ELT) reinforces this point and provides a vehicle for the CIO to connect the relevant dots between executive intentions and how IT can help turn those intentions into real-world accomplishments.

If you like coining new CxO titles, you might consider retitling the CIO to CCO (chief change officer) in recognition of this point.

---

* If they were, this book would be unnecessary.

Or not; name changes are often more of a distraction from what matters than anything else.

IT-Led Strategic Planning

Leadership. The word is inspiring. Even those who run away from decisions as if they were rabid wolverines utter it in reverential tones.

Back in the day, when computers were first escaping the confines of the accounting department, where they supported the general ledger, accounts payable, and payroll, IT led much of most companies' strategic planning.

The logic was uncomplicated. Everyone recognized that automating manual processes made processes cheaper and more reliable. Of course IT led the way.

Then it didn't. As related in chapter 1, IT relinquished its leadership role, instead taking on the guise of a merchant with wares to sell to its internal customers.

It's time for IT to take the lead again, not that the IT leadership of the future will look much like the IT leadership of days gone by.

Then, it was all obvious—choose the next process to automate, and automate it. Now we're in an era in which "Digital" has become both a noun and an imperative. The lead story in business strategic planning isn't process automation. It's capability. Here's why.

SWOT Spelled Backward Is . . .

The world is filled with strategic planning frameworks. One of the more popular is SWOT analysis, which stands for strengths, weaknesses, opportunities, and threats. As frameworks go, it has a lot going for it.

Except that it's spelled backward, because strengths and weaknesses are introspective, while everything that matters is extrospective. That is, no characteristic of any company is a strength or a weakness except in the context of externally facing threats and opportunities.

When CIOs reassume strategic planning leadership, they can do worse than pinning their fortunes on TOWS.

And in particular, on mapping potential new business capabilities, made possible by technological innovation, to the marketplace to understand whether or not they represent threats and opportunities.

That's threats *and* opportunities because the two are the same thing. A newly possible business capability is an opportunity if your company adopts it, and a threat if a competitor gets there first.

What's the CIO's role in all this? Recognizing how newly evolving and maturing technologies can create capabilities that might be relevant to the business model; discussing the possibilities with the executives and managers who might benefit from pursuing them; and, in larger companies, creating an organizational solution within IT for keeping track of all this.

As with so many other responsibilities, with growth comes the inability to do everything yourself.

Wait—Isn't Strategy the CEO's Job?

Answer #1: Yes.

If heading up strategic planning isn't in the CEO's job description, your company needs a different board of directors.

The first task of leadership is setting direction, and that's what strategic planning is all about.

What this means isn't that the CEO has the vision, translates it into the plan, and just tells everyone what the answer is.

What it does mean is that the CEO owns the strategic planning process, making sure it's fed by the best information and insights available.

That the CEO should lead the strategic planning *process* in no way diminishes the importance of IT taking the lead in establishing the strategic plan's *drivers*. The CEO builds strategic planning around a TOWS framework; the CIO plays a dominant role in establishing likely threats and opportunities; the CEO makes sure the ELT has the right conversations to assess their relevance to the business's future.

One more point: it's usually up to the CEO to draw the line that separates strategy from everyone on the ELT promoting their pet projects and horse-trading to end up with a consolidated list. A consolidated list is certainly a plan. It is, however, in no way a strategy.

Quite the opposite, it's the abdication of strategy—what a company does when nobody has a clear idea of what it should be doing.

Beyond this, refer back to chapter 3 (Fixing Agile) for thoughts on how to make strategy agile.

Answer #2: Yeah, but . . .

Another model that works transfers more responsibility for strategy to the CIO. In this model the CEO sets direction, goals, and priorities at the highest levels only. And, the CEO makes sure everyone in the ELT participates in the process, commits to the result, and recognizes their parts in making everything happen.

In this model the CIO (or CCO?) organizes and facilitates the planning process itself.

It's Okay to Lead

Leading strategic planning requires, obviously enough, leadership, which is to say getting the rest of the company to follow. This isn't a comfortable role for mainstream IT management, which is often so fearful someone will accuse it of squandering the company's scarce financial resources on technology for technology's sake that suggesting the company explore something new, different, and potentially useful is as intimidating as proposing a hefty assessment to pay for roof replacement at a townhouse owners association meeting.

So the starting point for technology leadership is, it's okay to lead. Not only is it okay, the business leaders we talk to are, for the most part, hungry for it. They want to have conversations about how they can use information technology to (1) get their work out the door more effectively, and (2) do something entirely new, different, innovative, and most important, profitable. Sadly, few in IT are equipped to have these conversations.

Properly led, IT is in the best position of any part of the business to have these conversations—to identify promising new technologies, incubate them, and help business leaders incorporate them into how we do things around here. It's technology leadership whether the technologies are new and emerging or just haven't been put to use in the business IT supports.

Sadly, many and perhaps most IT organizations aren't led this way. As evidence we offer the chief digital officer (CDO), a position defined as "what our CIO should be doing but isn't." Far too many IT leaders and professionals are shockingly unaware of new developments in their field, making IT as a whole unequipped for the conversations about capabilities businesses need to be engaging in, right now, all the time, for the foreseeable future.

To be fair, plenty of companies, and we're including many of those that undertake formal strategic planning exercises, aren't equipped for them either. Their strategic plans are tepid, fearful things. These companies don't think in terms of gaining marketplace advantage, beating the competition . . . you know, how to grow. Technology leadership can't happen in these companies, because no leadership can happen in these companies.

When decisions are rooted in fear, the outcome is stasis or retreat, not innovation and progress.

We hope your company is the other kind, poised to become an aggressive leader in its industry, with executives who want to try something more exciting than a stock buyback. They want to invest in competitive advantage. They just don't have better ideas about how to invest the company's spare cash.

All it might take is the suggestion from the CIO that (for example) adding intelligence to the company's products might make them more desirable; that social media intelligence might point the sales force to at-risk accounts, turning customer defections into increases in walletshare; that sales analytics will make both back orders and overstocks far less common than they are right now; that artificial intelligence can scan headlines to recognize events of interest to customers and shareholders, and even to craft and deliver messages—via Facebook, Twitter, or email—that cast the events in a favorable light. The possibilities aren't endless, but they are numerous.

Can One Person Keep Track of It All?

We're in the middle of an innovation wave, with new IT-driven capabilities appearing, evolving, and maturing so quickly it's hard to just keep track of what they are, let alone figure out their strategic implications and readiness for incorporation into the portfolio of business capabilities.

How to keep track depends on the size and scope of the business: entrepreneurships will have to make do with the CIO reading a handful of trusted sources, while small and medium-size businesses might create a position whose purpose is to read widely and champion pilot projects for the most promising possibilities.

Larger enterprises will probably have to establish a Research and Development–style office, perhaps integrated into the responsibilities of the enterprise architecture function.*

Whatever organizational solution you choose, the outcome should be to:

- Maintain a shortlist of promising new technologies—not promising in general, promising for your specific business.
- Perform impact analyses for each shortlist technology and keep them current, taking into account your industry, marketplace and position in it, brand and customer communication strategy, products and product strategies, and so on. Include a forecast of when each technology will be ripe for use.
- For each technology expected to be ripe within a year, develop an incubation and integration plan that includes first-business-use candidates and business cases, the logical IT (or, at times, non-IT) organizational home, and a TOWS impact analysis (threats, opportunities, weaknesses, strengths). Submit it to the project governance process.

---

* In case enterprise architecture isn't familiar territory, we provide a very quick sketch in chapter 7.

Governance Gateway

In chapter 5 we placed the company's strategic plan outside the scope of its Business Change Governance Council (BCGC). The reason is simple: the ELT has authority over the BCGC, not the other way around; likewise, the strategic plan takes precedence over departmental plans, except to the extent departmental plans are integral to achieving the strategy.

There is one force that has, or at least should have, even more authority than the ELT, and that's simple arithmetic. The company has fixed resources—fixed levels of staff, budget, time, and attention—that constrain its change bandwidth by limiting the total level of effort it can muster.

The BCGC is the keeper of the company's project master schedule. While it doesn't get a yea or nay vote on the collection of projects that make up the company's strategic plan, it does have an obligation to insert them all into the project master schedule, which exists to ensure the company doesn't pretend it has more effort to invest than is actually available.

While anyone might chair the BCGC, the CIO is the most likely candidate.

Which means the CIO is the one who has the uncomfortable responsibility of driving discussions regarding what inevitably happens while pouring the legendary ten pounds of project sugar into the proverbial five-pound resources bag.

To Sum Up

It's time for IT to resume its logical and necessary leadership role in the enterprise. While this might sound like a gratifying increase in the CIO's stature and importance, mostly it's going to mean neck out-sticking, political dexterity, and the need for deep insights into the marketplace and the company's current and potential place in it.

One more point: Rule #7 of the *KJR Manifesto* states, "Before you can be strategic you have to be competent."[1]

More than ever, businesses need IT to be in the lead. But to let IT take the lead, business leaders have to have confidence in its ability to deliver the goods.

If You Remember Nothing Else . . .

- SWOT—strengths, weaknesses, opportunities, and threats—is a useful strategic planning framework. Except that it's backward. The new SWOT is TOWS, because until leaders recognize the external threats and opportunities facing the business, they have no context for evaluating strengths and weaknesses.
- Before IT became a separate supplier and the rest of the business became its "internal customer," IT drove business strategy because automating the next manual process almost always made excellent sense. Now that IT is an equal partner and collaborator in designing and achieving business change, it's time for IT to resume its place at the planning table, because most threats and opportunities are the result of innovations in information technology, making IT the logical home for recognizing them and planning what to do about them.
- Strategic implementation projects aren't run through the change governance process. Because of their source, they are, by definition, approved. They are, however, incorporated into the project master schedule managed through the business-change governance process because the time, money, and staff required for their execution shouldn't be double-counted.

What You Can Do Right Now

- If you're the CEO, get in the habit of consulting with the CIO regarding IT-driven business threats and opportunities.
- If you're the CIO, make time to know how to answer the CEO's questions regarding IT-driven business threats and opportunities.
- If you're the CEO and the CIO doesn't step up, find a replacement capable of providing strategic IT leadership to the rest of the enterprise. Or, if you must, establish a CDO (chief digital officer) to provide IT leadership; but if you do, get ready for endless blamestorming and mutual finger-pointing.

7

The Seven Change Disciplines

We have a "strategic" plan. It's called doing things.
—Herb Kelleher, cofounder, Southwest Airlines

Interlude:
A Case of Good Intentions Not Being Good Enough

My employer, a regional storefront business, wanted to grow and transform, to emerge as a national and Digital player. Its executives wanted to expand their brand by offering new products and services beyond the company's existing regional locations.

The company's leadership and overall culture came from its long and successful history of transacting business in small customer-facing storefronts. Brand awareness and acceptance were consistent with the culture and solid among customers and employees. That heavily influenced internal operations, which was fully aligned with the small storefront operating model.

Which meant all of the above were badly misaligned with the vision of expanding into a new national and Digital business. The successful regional business practices, which were supposed to serve as the foundation, were, in fact, major barriers.

For example, the new business model called for centralizing product operations, bringing them in-house from external providers, and moving to a modernized internal service platform. From the outside it looked like a win all the way around—save costs, bring all serving in-house, and improve the customer experience for all consumer business products.

Some existing business products were already serviced internally, so integrating the rest didn't look like a huge challenge.

Except that they were supported by a decade-old software application. That's where the problems started. Operational processes were designed to cope with the limitations of the outdated application but had been "how we do things around here" for so long that few employees even recognized they were dealing with an accumulation of workarounds.

Thus, the reason why the new operations processes and supporting software were needed wasn't obvious to the employees on the front lines who had successfully provided top-notch service over the years. The storefront environment and accumulated workarounds were baked into their unconscious assumptions of how the world was supposed to be.

That was in business operations. In IT, the new application called for a move from Waterfall development to Agile, and a move from traditional application architecture to a services-oriented approach. As anyone who's worked in IT for any amount of time knows, these two moves are as culturally disruptive as the move from horses and buggies to automobiles.

And then . . . the transformation program teams lost the vision. Under cost and deadline pressures they turned into independent siloes of their own.

The result: what had started as an elegant federated data model ended up breaking information management and decision support. The architecture was focused on the technology modernization and not on end-to-end solutions designed with business operations and inte-gration in mind. The siloed technology solutions were stood up with limited success, but the end-to-end integration suffered.

The rollout delays were the least of the program's issues. Far worse was that without the end-to-end integration, the customer experience— the company's competitive advantage when it was a regional storefront-based company—was lost in the shuffle.

In the end, the biggest missing pieces were leadership, project manage-ment, and organizational change management. The company's leaders failed to keep the purpose of the changes in the forefront of everyone's minds; project management allowed individual teams to turn into uncooperative siloes; and the complete lack of organizational change management led to a widespread failure to see that the vision was broader than a technology modernization effort—it was a complete change in how the company defined its marketplace and competed in it.

Eventually, the program delivered: IT figured out more than Agile and modern application design—it bought into the idea that these weren't just IT projects. Business operations staff got the hang of pro-viding excellent service from a remote, centralized location, and decision-makers began to value the analytics they could get from a clean and consistent information repository.

But it sure would have been easier if these were skills and capabili-ties developed in the beginning of the change, instead of being lessons learned at the end of it.

Close your eyes and imagine a collection of colored balls con-nected by springs (or keep your eyes open and look at figure 4).* That's your organization—the one you're trying to change in some way, shape, or form.

---

* We know, we know. If you really closed your eyes, you wouldn't be able to read the rest of the instructions. Go with it.

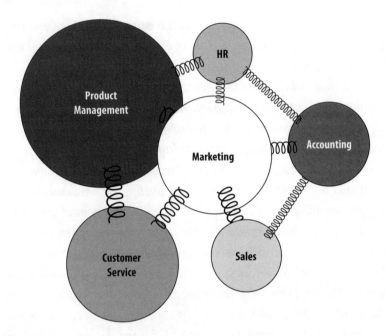

FIGURE 4 A metaphor for organizational change resistance.

Now imagine you grab one of the six balls and try to reposition it. Depending on how strong and tight the springs are, this shouldn't be all that difficult a task.

Repositioning that ball represents the change you're trying to achieve.

Now, let go of the ball. What happens? The entire organization returns to its pre-change state.

Changing an organization is hard. Not because "people just naturally resist change" (they don't) and not because figuring out what needs to change and what it should change to is hard (although it is).

It's hard because as someone once said, all organizations are perfectly designed to get the results they get.[1]

Which is to say, organizations, like most other systems, tend to evolve toward stable configurations. And it's the nature of stable that change gets resisted.

Most of this book has been about what needs to be *done* differently for intentional change to happen. But there's more to changing an organization than figuring out what has to be done differently and what it has to change into. This final chapter discusses, at a very high level, seven disciplines organizations need to master to make the intentional change, whatever it happens to be, both real and sustained. These disciplines are, in no particular order, leadership, business design, technical architecture management, application development or integration and configuration, organizational change management, implementation logistics, and project management.

One at a time . . .

Discipline #1: Leadership

As the old joke has it, there once was a man who had a dog with no legs. Every morning he had to take it out for a drag.

Leadership is the art of getting others to follow. For managers and executives who haven't mastered the art, getting the men and women in their organization to do what they're supposed to do is a tiresome drag, day after day after tedious day.

Leadership consists of eight responsibilities: setting direction, delegating, staffing, making decisions, motivating, managing team dynamics, establishing culture, and communicating, which includes listening, informing, persuading, and facilitating.[2]

In case the critical role leadership plays in achieving organizational change isn't clear, consider this: organizations are complicated, with a lot of moving parts that have, as noted above, coalesced into a stable configuration that works.

No change fiddles with just one of the moving parts. On top of this, every change destabilizes what used to be stable and requires a lot of hard work on everyone's part . . . not only the project team's members, but everyone . . . to make the change successful, deal with unexpected ripple effects, show the initiative needed to make sure everything continues to work during the interim, and, in the end, coalesce into a new configuration without jeopardizing the next organizational change, whatever it happens to be.

Employees who have to be dragged along aren't going to get the job done. "Leaders" who try to make the entire change happen without help aren't going to get the job done either.

Leadership is what changes the metaphor. With effective leadership, employees don't have to be dragged—they help pull the organization into its new configuration at maximum velocity without daily prompting.

And they go further, providing leadership themselves, figuring out the details and taking care of them without any need for managerial intervention.

One more point about leadership: as noted in the introduction, many organizations segregate themselves into a collection of squabbling siloes, or more accurately, castles with moats and drawbridges to control the flow of work in and out of them while maximizing the autonomy of each castle's owner to set direction and make decisions.

If executives and managers see themselves as leaders of their siloes, the company will be able to achieve intentional change within each silo just fine.

Strategic change? Probably not. To accomplish strategic change, those responsible for the various business units and functional areas . . . and beyond this, every manager, and beyond even this, every employee . . . have to see themselves as leaders of the whole company, focused now but not forever on a particular part of it, but motivated by the success of the enterprise as a whole.

Breaking down organizational siloes and keeping them broken down calls for unceasing leadership on the part of the company's top executives.

Discipline #2: Business Design

"Follow me!" we might imagine an inspiring leader crying to the troops.

"Where to?" we can easily envision the troops answering back.

"I don't know!" is a very bad answer.

Business design is about creating the answer—about designing the organizational change, whatever it is, completely enough so everyone involved can understand both the change itself and their role in it.

We covered business design in some depth in chapter 2. Business design excellence is self-evidently not just one of the critical organizational competencies needed for intentional change to take place but the first: if you don't know what the desired future state should look like, it will be awfully hard to make it happen.

Discipline #3: Technical Architecture Management

What technical architecture management is: an arcane but necessary IT function responsible for establishing the design and engineering guidelines needed so that the collection of applications, information repositories, and underlying plat-forms and infrastructure assemble logically and efficiently so as to support the organization's processes and practices.

Before we get started, readers who aren't in IT might be for-given for thinking they can skip this section. Doing so would be a mistake. Yes, technical architecture management is an ar-cane discipline, some of whose practitioners make it far more arcane than necessary.

But technical architecture management, done well, greatly re-duces the viscosity that's the all-too-common result of unman-aged, accidental technical architecture. Handled poorly, it ossifies IT, turning it into a major change bottleneck instead of a facilitator.

And, as handling technical architecture well requires invest-ment, arcane or not it's everyone's business, not just that of a handful of folks working in a back room churning out white papers.

Which gets us to the starting point: in our view the usual approaches for enterprise technical architecture management, built around documentation of the current state, a complete design of the desired future state, and a linear road map for moving from one to the other, share all the deficiencies associ-ated with Waterfall application development as explained in chapter 3.

Only they impose these deficiencies on a much wider scale.

An instruction manual on how to manage technical architec-ture is far beyond the scope of this book and would require an entire book to do the subject justice.

For this book's purposes, effective technical architecture management calls for IT to follow three key principles.

Principle #1: Each Application Obeys the Rules of Correct Data Design

If you aren't educated in the ways of information technologists, you don't need to know what this means. You just need to know that it matters, and why.

Why it matters: getting data into and out of an application with a poorly designed database can be miserably hard, and even harder to validate.

Bad data design is the problem behind a vendor being listed as, for example, "IT Catalysts, Inc.," "IT Catalysts Inc.," and "ITC," depending on the transaction record and who entered it.

It's the reason that a new address for one of your customers makes it into only some of your databases.

It's why you receive mailings from credit card companies offering you an introductory rate of 0% APR when you already have one of their credit cards, for which you're paying quite a bit more.

Increasingly, mining your data for meaning is critical to the success of your business. When data are inconsistent . . . when your databases disagree with each other . . . mining isn't just difficult. It's likely to give you an inaccurate answer to your questions.

Principle #2: Integration Engineering Is More Important Than Anything Else IT Applications Manages

As already discussed in chapter 3, most businesses, most of the time, buy or rent (that is, license or subscribe to) applications rather than building them internally.

But commercial applications come with their own databases and don't come with clean boundaries between themselves and the other commercial applications you have in play.

As a result, more than one application often manages overlapping information. A simple example: your CRM system keeps track of your customers; so does your accounts receivable system.

So, to some extent, do your social media analytics, if your business is such that you have customers who hang out on social media and you want to know what they're saying about you.

Now imagine your customer is a business, and it moves its headquarters. You have to make the change in two systems. Not a big deal, except that, as IT folks like to say, it doesn't scale.

Instead, you need software to keep the two systems' data synchronized.

Which might not seem like much of a problem. And it isn't when you have just two systems with overlapping data. Large, modern enterprises, though, might have two thousand or more applications with lots and lots of data overlap.

This leads to what some companies call their "interface spiderweb," others term "spaghetti," and those with the worst architectures and most honesty describe as their "IT hairball."

Whatever the name, we've seen IT shops that have more than a thousand separate custom data synchronization programs they have to run, in precise sequence, each and every night.

Imagine deploying a new application into an environment like this.

Replacing the hairball with clean and maintainable integration engineering is the most important, and strangely enough the most ignored, priority in technical architecture management. From the perspective of intentional business change it greatly reduces IT viscosity—that is, the sometimes enormous effort IT has to expend to make sure it hasn't broken anything when deploying and integrating the new software a business change depends on.

Principle #3: Every Project That Touches the Technical Architecture Leaves It in Better Shape Than It Found It

The problem IT has in improving the technical architecture is that doing so costs money and consumes capacity. Which means any project whose sole purpose is cleaning up the technical architecture has to compete for budget and staffing with other projects that require information technology but whose purpose is to achieve immediate intentional business change and improvement.

And so, with all the best of intentions and elegance of design, closing the gap separating the current state of things from the ideal future state the technical architecture folks have designed never happens.

The solution: don't even try to charter a separate technical architecture improvement roadmap. Instead, insist that every project that touches a company's portfolio of applications, databases, and underlying platforms and infrastructure move things in the right direction (right as defined by the technical architecture function) instead of doing more damage.

For example, we talked about the importance of integration engineering. A project that adheres to principle #3 would retire the point-to-point custom interfaces associated with the application being modified with connection to whatever integration engineering solution you've settled on.

Not your entire interface hairball. Just those interfaces the project in question has to deal with anyway.

Discipline #4: Application Development / Application Integration and Configuration

There is, we've all agreed, no such thing as an IT project. They're always about business change or what's the point? But while projects should always be about business change, it's equally true that most business-change projects place demands on a company's information technology, and in particular its portfolio of business applications.

At some stage of the proceedings, IT has to deliver new applications or make changes to the ones already in use so as to support the new way of doing things.

Probably tens of millions of how-to words have already been published on this subject. There would be little point in adding more. Two key points to complement them:

- **Implementing packages isn't the same as developing applications from scratch.** We covered this point in chapter 3 (Fixing Agile). Briefly, when you build from scratch you have to list the features you need and build them. When you buy a package they're already built. You need to converge them with your business processes.
- **The benefits from handling integration well last a very long time.** So does the pain of handling it wrong. We talked about this earlier in this chapter.

When it comes to achieving intentional business change, IT's goal is to avoid being the bottleneck. That's a key reason for mastering Agile techniques. It's a key reason for fixing your integration architecture.

More generally, it's a key reason for you to look at everything IT does and to remove everything you can that slows IT down.

Discipline #5: Organizational Change Management

Consider the curious case of *Who Moved My Cheese*,[3] probably the most popular and most misguided book ever published on the subject of change resistance. Its message to employees is (plot spoiler alert!) your cheese is going to move. Get over it.

That's to keep them from asking the logical and obvious questions: (1) Who decided to move the cheese, and (2) why are they messing with us for no good reason?

Real business changes have more moving parts to them than just a simple cheddar repositioning. With a more accurate metaphor, not only will employees find the cheese somewhere different from where they've learned to find it, but it's now Gouda instead of cheddar, and oh, by the way, employees now have an entirely different labyrinth to navigate before they can nibble on it.

And one more thing: there's a rumor going around that when someone does find the cheese, it will be nestled in the middle of a mousetrap.

People don't just naturally resist change. As evidence, we offer the smartphone. If that isn't evidence enough, while not everyone embraced the internet the moment the World Wide Web became real, it's pretty clear "people" didn't resist it.

Before that, the consumer electronics industry invented the compact disc, which was quite successful right up until streaming music services like Spotify also didn't meet with much resistance.

What confuses many business leaders on this subject is a straightforward distinction: while people don't just naturally resist change, they do just naturally resist change they expect will be bad for them. And as most business changes over the past four decades have had, as their centerpiece, one or more rounds of layoffs, employees' likely response to the next business-change or transformation program is going to be the expectation they'll be playing Whack-A-Mole once again, and they won't be the ones with the mallet.

If you still aren't convinced, imagine that instead of the change you're promoting and its consequences for employees, you're offering to give each of them a new car of their choosing, complete with five years of unlimited fuel, maintenance, and insurance. Think they'd all resist that change?

Neither do we.

To be fair, not everyone embraces every change immediately. Not everyone owns a smartphone; some of us are annoyed at the increasing number of things we used to do on personal computers that now require us to connect to the cloud; those of us with longer (but less reliable) memories often annoy those for whom our current events are taught as history in today's high schools by insisting that their use of social media is inferior to the pen pals a few of us had when we were young.

But few people reflexively resist any and all changes. Mostly, when employees (or anyone else) resist change they do so because they expect the change to be bad for them personally. Those who don't want smartphones might be insecure regarding their ability to figure them out; those who resist moving their data to the cloud might also be concerned about their ability to figure out how to make it all work or might have legitimate concerns about its security.

Even if the long-term result might be beneficial, the short-term transition to the changed situation might, as in knee replacement surgery, be painful enough to drive an unhappy reaction.

Which brings us to the first law of organizational change management: *to the extent possible, design every business change so it leaves employees better off than they were before the change happened.* That is, the individual employees, not just the organization but the men and women who do its work.

This in turn brings us to the first corollary of the first law: when you can't design a change so it leaves employees better off than they were before, at least design the change program so employees who support the change see a path to surviving it.

And it brings us to the second corollary of the first law: if a change is good for the organization but isn't good for anyone in the organization who's in a position to help drive the change . . . well, good luck with that.

This isn't the place for a complete guide to organizational change management. If you need one, check the endnote.[4] Somewhere between the first law and a complete methodology are a few critical tricks o' the trade:

- **Turn likely resisters into owners.** Figure out which stakeholders and stakeholder groups are likely to dislike what you have in mind, and find ways to involve them as active participants, even if you expect their participation to be annoying or disruptive. As LBJ reportedly said, better to have them in the tent spitting out than outside the tent spitting in.*
- **Find the systemic barriers.** Organizational resistance isn't limited to human beings who don't want it. Lots of structural factors can get in the way as well. Everything from facilities, to metrics, to your compensation system, to the chart of accounts Accounting uses to classify financial inflows and outflows . . . and let's not forget the organizational chart . . . can interfere as well. Figure out what

---

* LBJ's actual quote was a bit more colorful, but this is a family book.

they are and figure out how to change them or otherwise minimize their impact.

- **Provide training and support.** A common and reasonable reason employees resist some changes is that they're proficient in how to do their work right now. That makes them concerned they won't even be competent once the change takes place. The means for addressing this should be obvious.
- **You can't communicate too much.** When it comes to organizational change, ignorance isn't bliss. Quite the opposite, it's a source of anxiety. Let everyone know what's going on, not only when you launch and when you finish, but all along the way.

Discipline #6: Implementation Logistics

Organizational change isn't something you can just dump on an organization and hope those it's dumped on can figure out how to sort it out and make it work.

The usual approach is to start implementation with a proof of concept, pilot, or both.

Our advice about proofs of concept: don't bother.

Your average proof of concept addresses the simplest case that has to be dealt with, and that in a scaled-back fashion. It's a case the current systems and practices most likely handle just fine without the planned change.

A good proof of concept, in contrast, affects relatively few people, but otherwise includes most of the complexity of the actual rollout, only in miniature.

Which makes good proofs of concept indistinguishable from pilot implementations.

So start every business change implementation with a well-chosen pilot, giving it a lot of attention and support. Most important, view everything that goes wrong during the pilot as a positive event, as it's an opportunity to avoid similar problems when you start rolling out the change to the broader population.

Another recommendation: to the extent possible, turn change implementation into a machine—a team whose members know the drill because they've done it before and know they'll have to do it again.

This is especially true for changes that have to be deployed to a large number of different but similar locations, but it's also something to strive for in other implementations as well.

Whatever else, with most implementations the organization will find itself operating in a hybrid state for a significant period of time. This is usually as unavoidable as it is painful. The logistics plan has to prepare the organization so that everyone affected pitches in to make the transition work.

Discipline #7: Project Management

Projects are how change happens.

So if projects are how change happens, the critical role project management plays in achieving intentional change should be clear.

Businesses have no shortage of resources to draw on when it comes to project management methodologies.[5] Our purpose here isn't to provide a summary. It's to provide some guidance outside the methodologies themselves. Here's what you, as a change leader, need to know about project management:

- **What it is:** Project management is the means through which everyone involved in a project knows what they're supposed to do, when they're supposed to start doing it, and when it's due. And it deals with situations that cause deviations from the plan.

- **What it also is:** Project management is the discipline of figuring out the work that has to get done in order to produce whatever the project is supposed to produce. It ensures that if everyone does what they're supposed to do, when they're supposed to do it, delivering the results when they're due, the project will deliver what it was supposed to deliver.

  Very important: While project management is the discipline of figuring out the work that has to be done, this doesn't mean the project manager is the one who has to figure it all out. Quite the opposite. The project manager should, in consultation with SMEs, figure out the big blocks of work. Project team members should figure out the details, because (1) they're better equipped to do so, and (2) the plan is now their plan and not something the project manager has foisted on them.

- **One more thing it is:** Goaltending. That is, when it comes to project management, great ideas are the enemy of success. They're the source of "scope creep"—not the creep who moved my scope, but unfunded and unscheduled additions to what a project team has to deliver.

  The changes in scope themselves aren't the problem.* They become a problem when the team is supposed to just squeeze them in. A vital part of project management is enforcing the "Nothing Is Free" rule.

  A note about how Agile fits into this picture. The answer is, easily, because with most Agile variants, proposed

---

* The legendary example, from Defense contracting lore, has the project manager answering, "Oh, you wanted wings on that plane?"

changes in scope are expressed as features added to the project backlog. Once they've been added they're incorporated into the same priority-setting practices as everything else in the backlog.

- **What it's the opposite of:** Everyone waking up in the morning trying to figure out how to move the project forward today.
- **Culture leads again:** Even the best project managers can't succeed in organizations that lack a culture of project management, which is to say a culture that recognizes the importance of disciplined planning and execution and supports it.

Some companies still view project management as a "bridge position"—a way to give promising individual contributors some management experience, following which, should they succeed, they're promoted to management. This is a bad idea for two reasons. First, projects are led by inexperienced project managers. Second, this practice prevents the organization from ever accumulating a cadre of experienced project managers.

As project management is one of the seven critical disciplines for achieving organizational competence at intentional change, we hope it's obvious why the project manager role is too important to be relegated to being a mere bridge to something else.

Quite the opposite: as one of the most important management roles in the organization, it warrants its own career path.

One More Thing . . .

Mastery of the seven disciplines—of leadership, business design, technical architecture management, application development or integration and configuration, organizational change management, implementation logistics, and project management—is a necessary condition for achieving organizational competence at intentional organizational change.

But there's a risk. As with anything else that appears on the organizational chart, those who provide these capabilities to the enterprise can easily turn them into bureaucracies and organizational siloes, protective of turf and squabbling over budget and influence.

To make it all work, mastery in isolation isn't enough—the seven disciplines have to come together as an integrated whole, whose practitioners actively collaborate to make intentional change happen.

If You Remember Nothing Else . . .

Businesses that have mastered the art of intentional business change have achieved competence at seven core disciplines:

- Leadership
- Business design
- Technical architecture management
- Application development or integration and configuration
- Organizational change management
- Implementation logistics
- Project management

What You Can Do Right Now

- Assess the enterprise's overall organizational competence in each of the seven disciplines. If the objectivity and expertise for doing so don't exist inside the company, you might need to engage an outside consultant for this purpose.
- Give whoever is currently responsible for each of the seven disciplines a copy of this book (sorry!) to start them moving

in the right direction and to encourage them to start collaborating on making the enterprise adept at intentional
business change.

- Don't try to fix everything at once and with the same
executive focus and intensity. Apply the principles of this
book for instituting a business that embraces the principles
of this book.

Epilogue

A Few Last Words

*Our ability to adapt is amazing. Our ability
to change isn't quite as spectacular.*
—Lisa Lutz

There you have it. To make change happen, all you gotta do is . . . well, first of all you have to ignore any and all advisers who start their pitch with "All you gotta do is . . ."

As you can see if you've made it through, "all you gotta do" is a dangerous message. Modern businesses are complex. They're complex externally, where their decisions about products, services, pricing, customer targeting, selection of vendors and partners, and branding and messaging all have to come together.

They're complex internally, too, as employees and contractors:

- Participate in business processes using the tools IT provides.
- Receive work direction from whoever they currently report to in a world of continuous reorganizations.

- Make decisions in large part as a consequence of the un-conscious assumptions and habits reinforced by the business culture they're immersed in.

While there are any number of "all you gotta do" panaceas out there, promising to make it all simple, when it comes to organizational change, simplicity is wishful thinking.

Trend-spotting readers might wonder why we didn't spend much time or place much emphasis on Digital, either as a noun or as an adjective. It's a reasonable question, given that while "Digital" is used far more often than it's defined, it's nonetheless quite important to the future of most businesses.

Ignoring the multitude of definitions, what makes Digitality particularly important is its intense focus on increasing revenue rather than on decreasing costs as strategy's primary imperative, and, as a consequence, its equally intense focus on enhancing the customer experience rather than on increasing internal process efficiencies even when it's to the customer's detriment.

So while we didn't intersperse "Digital" throughout the book you just read, we trust you'll recognize that this sort of thinking perfuses it. We didn't emphasize things Digital for a simple reason: all Digital business changes are business changes, but not all business changes are Digital.

Both of us have led business changes of various sizes and scopes, participated in more, and been observers of no small number of them besides. Some of the changes were more successful than others. Which were the more important sources of insights for this book is open to question.

What is fair to say is that making change happen is hard. It's hard the way neurosurgery is hard; it's also hard the way digging a ditch is hard.

A common metaphor is that it's like replacing an engine on an airplane while the airplane is in flight. It's a reasonable description of most of the changes we've been involved in.

When leading change, most days felt like we were taking the proverbial two steps forward to one step back, but there was no shortage of days that reversed the numbers.

Not to mention the many occasions in which we heard participants take great pride in a recent accomplishment, only to discover their "accomplishment" was based on a completely flawed understanding of what we had in mind. Their flawed understandings usually resulted from a cultural mismatch, which is to say their perceptions about what we had in mind were as much a reflection of their own unconscious preconceptions as of our explanations.

This book is a prime example. As we've discussed the premise with colleagues, friends, and clients we've found ourselves colliding with preconceptions over and over again: "There's no such thing as an IT project," we explain, continuing, "It's always about business change or there's no point to it."

"We're way ahead of you," we hear in response. "We make sure every project we undertake delivers business benefit."

We hope we've made the point clearly and with no more repetition than necessary that business benefit isn't the same as intentional business change. Yes, of course the intended business changes should be beneficial. But no, if the project is defined in terms of delivering software, using the software to make business change happen will be Someone Else's Problem.

And that Someone might not see it that way.

We have, on occasion, found ourselves explaining the challenges of management, leadership, and making change happen to technical professionals promoted into leadership, management, and project management roles.

Our metaphor for the situations they've found themselves in is that they should view the organization they manage as a computer that has a glitchy operating system and buggy programming language.

They still have to program the sucker, which means finding workarounds for all the glitches and bugs.

Only when it comes to an organization, the glitches and bugs are the idiosyncrasies of those pesky human beings who populate it.

The other point we make is that we recognize the satisfaction that comes from writing a computer program that does something useful.

It's a satisfaction that pales in comparison with successfully achieving intentional organizational change.

Which brings us back to the design and implementation of a swing—the metaphor we used to start this handbook of organizational change.

We hope you see the differences. Not only the differences between all the miscommunications that can happen when every step of swing design doesn't keep why anyone might want a swing firmly in mind. But also the differences between delivering a swing, no matter how well designed and constructed, and your kids and their friends actually playing on it and around it.

Now that you understand the differences, you're ready to make intentional business change happen.

Notes

Prologue

1. Source unknown, which is a shame, given the cartoon's utter perfection.

Chapter 2

1. Wikipedia, s.v. "Philip B. Crosby," last modified December 26, 2018, https://en.wikipedia.org/wiki/Philip_B._Crosby.
2. Not original, but original author unknown.
3. Eliyahu Goldratt, *Theory of Constraints* (Great Barrington, MA: North River Press, 1990).
4. Paco Underhill, *Why We Buy: The Science of Shopping*, rev. ed. (New York: Simon and Schuster, 2008).
5. Jim Collins, *Good to Great: Why Some Companies Make the Leap . . . and Others Don't* (New York: HarperCollins, 2011).
6. Usually attributed to Benjamin Disraeli, but according to the Phrase Finder (www.phrases.org.uk) it appears the credit more likely belongs to Arthur James Balfour, 1st Earl of Balfour; see https://www.phrases.org.uk/meanings/lies-damned-lies-and-statistics.html.

Chapter 3

1. "Chaos Resolution by Agile versus Waterfall," Standish Group, 2015, https://platinumedge.com/sites/default/files/public/Standish-Group-2015-Agile-v-Waterfall.pdf.

2. "Someone a long time ago" means we weren't able to track down the original research.

3. Bob Lewis and Scott Lee, *The Cognitive Enterprise* (Tampa, FL: Meghan-Kiffer Press, 2015).

4. *Manifesto for Agile Software Development*, Agile Alliance, http://agilemanifesto.org/.

Chapter 4

1. Sun Tzu, *The Art of War* (original publisher and publication date unknown).

Chapter 6

1. Bob Lewis, *Keep the Joint Running: A Manifesto for 21st Century Information Technology* (Eden Prairie, MN: IS Survivor Publishing, 2009).

Chapter 7

1. Usually ascribed to W. Edwards Deming, but likely originating with his colleague Arthur Jones. "Quote by W. Edwards Deming," Deming Institute, 2019, http://quotes.deming.org/authors/W._Edwards_Deming/quote/10141.

2. Bob Lewis, *Leading IT: (Still) the Toughest Job in the World,* 2nd ed. (Eden Prairie, MN: IS Survivor Publishing, 2011).

3. Spencer Johnson, *Who Moved My Cheese* (New York: G. P. Putnam's Sons, 1998).

4. Bob Lewis, *Bare Bones Change Management* (Eden Prairie, MN: IS Survivor Publishing, 2010).

5. The best: Bob Lewis, *Bare Bones Project Management* (Eden Prairie, MN: IS Survivor Publishing, 2011). Second-best: Eliyahu Goldratt, *Critical Chain* (Great Barrington, MA: North River Press, 2002).

Glossary

Agile (n): (1) A family of application development methodologies characterized by high levels of direct, informal interaction between developers and end users; and by adding capabilities to software iteratively and incrementally instead of trying to design the perfect solution before programming starts. "Agile" should be an adjective, but that battle was lost a long time ago. Contrast with "Waterfall." (2) The application of iteration and incrementalism to business strategy.

ATDD (Acceptance-Test-Driven Development) (n): An Agile variant in which the development team first defines the acceptance tests the application must pass, then writes software that can pass them. See also "CRP."

backlog (n): In Agile, the repository of all epics, features, and user stories.

black-box analysis (n): A view of a business function that describes only outputs and inputs, without any explanation of how it transforms its inputs into outputs.

blamestorming (n): A popular alternative to root cause analysis that focuses on who to blame for a problem instead of performing the more difficult and expensive process of root cause analysis. *Example result: "We have a blame-oriented culture and it's your fault."*

Business Change Governance Council (n): As proposed in this book, the organization responsible for effective change governance. See also "change governance" and "IT Steering Committee."

business function optimization (n): The practices needed to determine which of the six dimensions of optimization are most important for a given business function, and for taking the steps necessary for making the function as effective as possible with respect to the most important dimensions of optimization.

business process reengineering (n): A process design methodology that imagines the existing process is pretty much worthless and it's time to start with a clean sheet of paper (or, more accurately, a blank Visio page). It can be useful when integrating multiple organizations. Otherwise, it's just one more process design religion that ends up fixing what's broken by breaking what's fixed. See also "Lean," "Six Sigma," and "Theory of Constraints."

BusOps (pronounced 'Biz-Ops) (n): A term introduced in this book that describes a new relationship between IT Operations and Business Operations in which the two are harmonized so they collaboratively "keep the joint running" and make sure it's running as well as possible. See also "IT Operations."

change governance (n): The set of practices proposed in this book for deciding which business-change efforts should be undertaken. It's the art of helping good ideas succeed. See also "Business Change Governance Council" and "IT Steering Committee."

change steward (n): The term, introduced in this book, describing the role that corresponds to an Agile product owner when Agile has been adjusted to organize business change instead of just software delivery.

configuration (n): A way for companies to make COTS applications do what they need them to do by making use of built-in capabilities for adding data fields and functionality. Compare with "customization."

consultant (n): (1) An outside expert brought in to an organization to provide objective and unbiased advice and guidance. (2) A self-described expert whose primary skill is fixing what's broken by breaking what's fixed.

consumers (n): Those who make use of a company's products and services. Contrast with "customer" and "wallet."

COTS (commercial off-the-shelf software) (n): Prepackaged software an individual or business can license for use. COTS is typically installed on a company's own servers and is usually licensed as a combination of a core license plus additional per-client fees. Yes, it should be COTSS. But it isn't, and there's nothing any of us can do about it.

CRM (customer relationship management) (n): A suite of business applications that supports all interactions with real, paying customers along with all of the information known about them.

CRP (conference room pilot) (n): An Agile variant optimized for implementing COTS and SaaS applications. Similar to ATDD but less formal.

culture of honest inquiry (n): A corporate culture in which everyone embraces the conclusions that result from where the best evidence and soundest logic lead. Those who lead businesses that have this culture understand that their gut is for digesting food—their brain is where thinking takes place.

customer (n): The person who makes a buying decision, along with those who strongly influence it. Contrast with "consumers" and "wallet."

customization (n): A horrible and shortsighted method companies use to make COTS applications do what they need. Companies that practice customization modify data structures and core application code, which makes applying software updates expensive, problematic, and generally a miserable experience.

cycle time (n): The average time needed to turn one business function input into an output.

data lake (n): A highly scalable repository of raw, unfiltered data intended for exploration and analysis by data scientists who understand the hazards and pitfalls of performing statistics on data not collected for that purpose. Compare with "data warehouse."

data warehouse (n): A highly scalable repository of carefully filtered, scrubbed, and organized data intended for use by just about anyone who needs to analyze the data the data warehouse manages.

decision support (n): The practices, information repositories, and analytical applications needed for a business to inform its decision-making with the best available evidence.

DevOps (n): An Agile variant whose proponents think they invented collaboration between IT Apps and IT Ops. Originally, DevOps meant including someone from IT Ops on Application Development teams. What makes DevOps interesting is its insistence on automating everything that can be automated, and its diligent insistence that the software being developed is always in a deployable state. The meaning behind the common description of the DevOps methodology is that it practices continuous integration and continuous deployment.

enlightened (participle): A term that describes all executives, managers, and staff who have read this book and accepted the principles it propounds.

epic (n): In Agile, a high-level description of something the system should do, described in very broad terms. *Example: create and edit documents.* Compare with "feature" and "user story."

ERP (enterprise resource planning) system (n): (1) Proper usage—a business application suite designed to support enterprise resource planning practices. (2) Usual usage—as nobody knows how to go about "enterprise resource planning," ERP is used to describe a suite of applications that supports a business's core internal processes and practices such as core accounting, supply chain management, and order management.

excellence (n): For a given business function, its ability to adapt, tailor, customize, or otherwise create value irrespective of quality.
    Example: A Lamborghini that leaks oil all over the garage floor is low quality but still excellent, as contrasted with a Dodge Neon that

starts, accelerates, brakes, and stops every time you drive it, even though it's boring.

experience engineering (n): The practices required to make the experiences users and customers have as un-irritating as possible.

feature (n): In Agile, the capabilities needed for an epic to be satisfied. *Example: enter text, format text, store documents as files.* Compare with "epic" and "user story."

fixed costs (n): For a given business function, the costs incurred before processing any work. The metaphorical "cost of turning on the lights."

illuminated IT (n): A complimentary term, used by enlightened IT management to refer to business DIY application development and implementations that are supported and assisted by IT so as to avoid risks and potential compliance issues.

incremental cost (n): For a given business function, the additional cost incurred by processing one more item.

incrementalism (n): A business-change philosophy built around evolutionary principles—that small changes can accumulate into transformative results with lower costs, less risk, and more rapid delivery of value than attempting to craft a perfect design from scratch.

information technology (IT) (n): (1) Computers and associated technical items, and what they can be and are used for. (2) The organization within the business responsible for its computers, what they're used for, and what they could be used for but aren't yet. See also "IT Applications" and "IT Operations."

invisibility index (n): The only metric that matters in assessing the performance of IT Operations. The invisibility index determines how visible IT Operations was during a reporting period, on the grounds that the only time anyone notices IT Operations is when something goes wrong, so that perfect invisibility is the best possible rating.

IT Applications (Apps) (n): The part of an IT organization responsible for developing, enhancing, and maintaining new software; or licensing, installing, integrating, and configuring commercially available software. IT Applications succeeds by changing yesterday into tomorrow.

IT Operations (Ops) (n): The part of an IT organization responsible for making sure all of the business's information technology is running properly. IT Operations succeeds by making sure tomorrow looks as much like yesterday as possible.

IT Steering Committee (n): An organization devoted to coordinating the efforts of each business function to get as much of IT's time, attention, and budget as possible. (2) An organization devoted to preventing bad ideas from receiving any further consideration. See also "Business Change Governance Council" and "change governance."

Kanban (n): (1) A practice within Lean in which those who perform work receive their assignments from a queue of work to be done. (2) An Agile variant that follows the Lean practice of the same name. Kanban application development is less rigid and more flexible than Scrum, letting everyone involved concentrate on getting useful work done.

Lean (n): A process design methodology that, when properly applied, reduces waste by optimizing work-in-progress queues.

When improperly applied it's just one more process design religion that ends up fixing what's broken by breaking what's fixed.

legacy system (n): A puzzling term that should mean a valuable system current practitioners inherit from their predecessors, but that in actual use means a business application akin to a boat anchored and becalmed in the middle of the Sargasso Sea: it floats, but its passengers and crew can neither leave nor make progress.

marketshare (n): The fraction of all spending for a product or service category captured by a business.

methodology (n): (1) Proper usage—the study of methods. Nobody ever uses this definition. (2) Common usage—a method. (3) Accurate usage—a business "religion," applied to situations whether or not it fits those situations.

mindshare (n): The extent to which potential customers, faced with making a buying decision, think of the business in question instead of its competitors.

multivariate (n): (1) In statistics, analyses used when multiple causal factors can influence a given effect. (2) In this book, situations in which multiple factors all influence a given effect, resulting in a high degree of difficulty determining which one or ones are responsible for a given outcome.

pain point (n): (1) Usual usage—anything anyone doesn't like. (2) As used by the enlightened readers of this book, for a given business function, one of the six dimensions of process optimization that's both highly ranked and unsatisfactory.

persona management (n): Categorization of customers and consumers into typical groups that experience your company, characterizing them by way of an evocative name as a convenient shorthand. *Examples: "Demanding Dan" describes difficult but highly profitable customers. "Agreeable Anne" covers customers who accept whatever experience they happen to get without complaint . . . until they get on social media.*

practice (n): A way to organize a business function so that success depends on the knowledge, skills, experience, and overall judgment of the practitioner. Practices are typically optimized for fixed cost, cycle time, and excellence. See also "process."

pretzel logic (n): Business workflows that follow baroque and convoluted paths. The inefficiencies associated with pretzel logic are usually a consequence of being forced to use outmoded or poorly constructed supporting business applications.

process (n): A way of organizing a business function into real or metaphorical assembly lines, where success is the result of following the right steps in the right sequence. Processes are "designed by geniuses to be run by idiots" and are typically optimized for incremental cost, throughput, and quality. See also "practice."

process bypass process (n): Your loyal authors' term for processes whose purpose is to allow employees, faced with situations the official process doesn't fit, to bypass the process without process managers completely losing control of the situation.

quality (n): (1) Usual usage—good, or what makes something good. (2) As proposed by Philip Crosby and used in this book, adherence to specifications and the consequent reduction in defects.

root cause analysis (n): When a business outcome isn't what's desired, or an undesirable situation arises, the painstaking process of determining what fixable characteristics of the organization either caused the outcome or situation or failed to prevent it.

Rummler-Brache diagram (n): See "swim-lane diagram."

SaaS (Software as a Service) (n): A form of COTS that's made available through the internet instead of being installed on a company's own computers. SaaS is typically licensed on a pay-as-you-go basis.

SAFe (Scaled Agile Framework) (n): An attempt to address the strategy-to-action impedance mismatch by adding structure and rigor to Agile's informality. That is, it's an application development methodology that grafts a thin Agile skin on a Waterfall heart.

Scrum (n): The most popular Agile variant, probably because it's the most highly structured, and therefore most palatable to project managers accustomed to Waterfall's tight control over application development.

shadow IT (n): A derogatory term used by IT management to refer to business DIY application development and implementations, although not as derogatory as "rogue IT." Contrast with "illuminated IT."

shelfware (n): Anything, but especially software and consultant recommendations, that's been bought, paid for, and made available but that nobody actually uses. *Example: most management books, but not this one.*

six dimensions of optimization (n): Fixed cost, incremental cost, cycle time, throughput, quality, and excellence (definitions provided).

Six Sigma (n): A process design methodology that, when properly applied, reduces variability among process outputs (products), thereby reducing the number of defective products produced. When improperly applied it's just one more process design religion that ends up fixing what's broken by breaking what's fixed.

SLA (service level agreement) (n): (1) A term in contracts negotiated between service providers and their customers that specifies a minimum level of service and how often the service provider must reach or exceed it. (2) A formal contract negotiated between IT managers who think they're running a business that supplies information technology and related services, and their so-called internal customers. Like a true contract, it places the relationship between the two parties at arm's length. Unlike a true contract, it provides no remedies for nonperformance because don't be ridiculous.

SME (subject matter expert) (n): Someone who knows more than we do about the subject at hand.

SSC ratio (n): A metric developed by Bob Lewis and Scott Lee that compares the time a business change will be useful in production (numerator) with the time needed to accomplish the change (denominator). The SSC ratio explains the impact of change acceleration: think of it as how long food you grow and prepare lasts before it becomes too moldy to eat.

The acceleration of change means the time-to-mold metric is getting shorter, but the time needed to plant, grow, harvest, and cook isn't.

strategy-to-action impedance mismatch (n): What happens when Waterfall-oriented strategic planning collides with iterative and incremental Agile implementation methodologies.

swim-lane diagram (n): A type of flowchart that's particularly useful for white-box analysis. It lays the process flow in a grid, with rows representing actors and columns representing steps, presenting both process flow and responsibilities in a single view.

Swim-lane diagrams were invented by Geary Rummler and Alan Brache, a fact known by a small fraction of the consultants that use them.

SWOT (strengths, weaknesses, opportunities, threats) (n): A useful but inverted framework for business strategic planning. Compare with "TOWS."

technical architecture management (n): (1) An arcane but necessary IT function responsible for establishing the design and engineering guidelines needed so that the collection of applications, information repositories, and underlying platforms and infrastructure assemble logically and efficiently so as to support the organization's processes and practices. (2) An ivory-tower white-paper factory.

Theory of Constraints (n): A process design methodology that, when properly applied, improves process capacity and throughput by speeding up process bottlenecks (constraints). When improperly applied it's just one more process design religion that ends up fixing what's broken by breaking what's fixed.

throughput (n): A business function's realized capacity—the number of inputs it can turn into outputs in a given unit of time.

touchpoint (n): As used by the enlightened (those who have read this book), a touchpoint is the intersection of a process step and an interaction channel. Placing an order on a company's eCommerce site might be a touchpoint; placing the same order through the company's call center might be another; using online chat for after-purchase support might be a third.

TOWS (threats, opportunities, weaknesses, strengths) (n): A useful strategic planning framework introduced in this book. It's superior to the better-known SWOT in being externally focused: weaknesses and strengths have no meaning except to the extent they affect the organization's ability to pursue opportunities and counter threats.

user story (n): What Agile has instead of traditional requirements. It describes something specific a user wants to do. *Example, standard form: As a document creator I want to be able to define styles that describe all formatting for selected text, so I can easily format document parts consistently.* See also "epic" and "feature."

value lever analysis (n): An approach to business modeling that tries to take into account the multiple linkages and connections between actions a company can take and their bottom-line impact.

wallet (n): One of three "customer" roles; the wallet is the person or group that provides the money used to buy a product or service. Contrast with "customer" and "consumers."

walletshare (n): The fraction of everything a customer spends for a product or service category provided by a business that's spent with that business.

Waterfall methodologies (n): A family of project management methodologies that divide the work into phases, where once you finish a phase—once you go over the Waterfall—it's expensive and disruptive to go back to the top. In application development, methodologies that attempt to fully specify the solution before it's possible to fully understand the problem.

Alternate etymology: it's called Waterfall because once you go over it, you crash onto the pile of rocks at the bottom and aren't likely to survive the experience.

white-box analysis (n): A view of a business function that describes how it transforms its inputs into outputs. White-box analysis usually consists of a nested set of flowcharts or swim-lane diagrams.

Acknowledgments

It starts at home: we're grateful to our wives, Sharon and Jean, for putting up with not only the time commitment but also the usual gnashing of teeth and incoherent grumbling that accompany the process of figuring out what we want to say, how we want to say it, and in what order so it makes some semblance of sense.

As someone who has read or plans to read the book in front of you, you should be even more appreciative than we are of, in alphabetical order (first name), Andy Gebhard, Anita Cassidy, Chad Hagedorn, Harold Knutson, Jonathon Kass, Kelly Williams, Max Fritzler, Mike Benz, and Tony Carangelo.

These fine folks, who live the principles we wrote about here, patiently reviewed each chapter as we produced it to help us make sure we explained our thinking effectively, point out what we'd missed entirely, and, on occasion, diplomatically suggest that some of our dulcet prose should be simply excised from the text.

This work is immeasurably better for their participation, and if you liked reading it, they deserve quite a lot of the credit.

Speaking of reviewers, we need to thank Mike McNair and Candace Sinclair, the professional readers to whom our editors assigned the thankless task of reading our completed manuscript and suggesting ways to improve the overall narrative. You have them to thank for the chapter summaries and action plans. Also, to the extent the overall flow of the book makes sense, you have them to thank for much of that too.

From Dave, a shout-out to Bill Domings, now retired and formerly of McDonald Consulting. Bill was instrumental in showing Dave the power of multifunctional teams and how to create high-performing customer-focused organizations. Bill never stopped telling me the importance of managing the business versus processing the transactions. Bill left his mark on many organizations and people throughout his career.

If the book publishing industry gave awards for acts of everyday patience, one would surely go to Charlotte Ashcroft, our editor at Berrett-Koehler. She gets the credit for calmly responding to our occasional cries of "We have to do what?!?!?" as she walked us through the ins and outs of our responsibilities following delivery of what, in our eyes, was already a perfect manuscript.

Then there's Jeevan Sivasubramaniam, Berrett-Koehler's managing editor and one of Bob's long-standing friends-he's-never-met. Over many years he and Bob agreed they needed to do a book together. And over that same span of years he patiently (patience appears to be built into Berrett-Koehler's business culture) explained that no, the concept Bob thought was so deeply important wouldn't turn into a book that would, for example, sell.

And then, when Bob suggested this one, he agreed that this time, perhaps, it would.

Speaking of friends we've never met, there's the *Keep the Joint Running* community. These are the fine folks who read Bob's weekly e-letter and, in their online comments and email correspondence, have, over the past sixteen-plus years, helped refine and improve the idea that no, there really is no such thing as an IT project, an idea first introduced in *KJR's* predecessor, *InfoWorld's* "IS Survival Guide," back in 2002 (https://issurvivor.com/2002/04/01/the-new-world-of-prototyping-first-appeared-in-infoworld/).

Our thanks to one and all.

Index

About the Authors

Before becoming a recognized industry pundit (RIP), **Bob Lewis** amassed five years of experience researching the behavior of electric fish and, more broadly, sociobiology, followed by ten years of experience applying the knowledge he'd gained by working in what's often called the "real world." This took the form of a variety of staff, management, and leadership roles, including stints in information technology, product development, manufacturing analysis, and modeling.

Twenty-two years of writing and consulting followed, resulting in quite a lot of direct and indirect knowledge and experience about what makes some organizations more effective than others, and how to get from here to there without pretending people will behave how we'd like them to behave, just because we'd like them to.

When *InfoWorld* was a print publication, Bob was one of its most popular columnists; his "IS Survival Guide" reached an estimated 250,000 readers per week and received a silver medal from the ASBPE, West Coast chapter. Counting this, his *InfoWorld* blog *Advice Line* (which received a bronze medal from the same organization), various feature stories published in *InfoWorld* and *CIO*, and his own *Keep the Joint Running* column/blog, he's published a dozen books (including this one) and more than 1,600 columns on business strategy, leadership, organizational effectiveness, and how to integrate information technology into all of the above and more.

Dave Kaiser started his IT career at the ripe age of nineteen at industry-leading Control Data, which ceased to exist just a few years later. He has twenty-three years of experience in IT leadership and is currently the CIO of a Midwest-based insurance company.

His focus has been leading people, establishing a healthy corporate culture, process engineering, software development, and providing superior customer experiences through information technology. He is a strong believer in using smaller teams to deliver faster results that also happen to be on-budget, something that somehow seems unusual in today's corporate world.

Dave has presented at numerous industry conferences in a wide range of subjects, from software development trends to predictive analytics. He is also the cofounder and past CEO of the charitable SFM Foundation, which is in its eleventh year of granting scholarships to the children of workers severely injured in the workplace.

Dear reader,

Thank you for picking up this book and welcome to the worldwide BK community! You're joining a special group of people who have come together to create positive change in their lives, organizations, and communities.

What's BK all about?

Our mission is to connect people and ideas to create a world that works for all.

Why? Our communities, organizations, and lives get bogged down by old paradigms of self-interest, exclusion, hierarchy, and privilege. But we believe that can change. That's why we seek the leading experts on these challenges—and share their actionable ideas with you.

A welcome gift

To help you get started, we'd like to offer you a **free copy** of one of our bestselling ebooks:

www.bkconnection.com/welcome

When you claim your **free ebook**, you'll also be subscribed to our blog.

Our freshest insights

Access the best new tools and ideas for leaders at all levels on our blog at ideas.bkconnection.com.

Sincerely,

Your friends at Berrett-Koehler

MIX
Paper from
responsible sources
FSC® C008955

Certified

Corporation